Allies in Air Power

Allies in Air Power

A History of Multinational Air Operations

Edited by Steven Paget

UNIVERSITY PRESS OF KENTUCKY

Scholarly publisher for the Commonwealth,
serving Bellarmine University, Berea College, Centre
College of Kentucky, Eastern Kentucky University,
The Filson Historical Society, Georgetown College,
Kentucky Historical Society, Kentucky State University,
Morehead State University, Murray State University,
Northern Kentucky University, Transylvania University,
University of Kentucky, University of Louisville,
and Western Kentucky University.
All rights reserved.

Chapter 12 © Her Majesty the Queen in Right of Canada 2020.

Editorial and Sales Offices: The University Press of Kentucky
663 South Limestone Street, Lexington, Kentucky 40508-4008
www.kentuckypress.com

Cataloging-in-Publication data is available from the Library of Congress.

ISBN 978-0-8131-8032-8 (hardcover : alk. paper)
ISBN 978-0-8131-8033-5 (pdf)
ISBN 978-0-8131-8034-2 (epub)

This book is printed on acid-free paper meeting
the requirements of the American National Standard
for Permanence in Paper for Printed Library Materials.

Manufactured in the United States of America

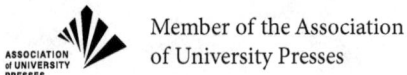

Member of the Association
of University Presses

Contents

Introduction

Multinational Air Power: The Context

Steven Paget

Multinational military operations predate air power significantly, but the trajectories of both are now intertwined as contemporary air operations are generally characterized by international cooperation. This fact has been encapsulated in the doctrines of air forces around the world. The United Kingdom's Air and Space Power doctrine has outlined: "Our dependence on multinational cooperation means we must take every opportunity to promote interoperability and engage with a broad range of potential partners. These range from the most technologically sophisticated air forces to less familiar and capable but potentially likely partners."[1] The Royal Australian Air Force's *Air Power Manual* notes succinctly: "In the prevailing geostrategic environment, there is a high likelihood that any major operation involving Air Force will be conducted within a coalition."[2]

Unilateral air operations are becoming increasingly rare for most air forces as "the norm of fighting alongside others is now deeply entrenched."[3] Frequency should not be equated with simplicity, however, because multinational operations have "at all times been hallmarked by significant challenges."[4] Indeed, although multinational air operations have been conducted since the birth of flight, barriers to integration have persisted into the contemporary era. The speed with which air power has progressed and its status as "a dominant, if not the dominant military technology over the last century" have served only to heighten the challenges experienced when air power is conducted in a multinational environment.[5] Contributions from different nations provide many benefits—from niche capabilities to political credibility—but they also present a number of challenges. Issues such as command and control, communications, equipment standardization, intelligence, logistics, planning, rules of engagement, training, as well as tactics, techniques, and procedures require consideration. Cultural factors, in addition, can present a number of challenges, particularly when language barriers are involved.[6] Air forces have confronted this requirement to balance the challenges and opportunities presented by multinational operations from the dawn of air power to the present day.

The actual conduct of operations is only the culmination of multinational cooperation and frequently represents the fruition of peacetime interaction and previous experience. Wing Commander André Adamson, Royal Air Force, and Colonel Peter Goldfein, US Air Force, have emphasized: "Operations act as a catalyst to integration (through sheer necessity), but difficulties that emerge during complex multinational operations point to the need to pre-empt those frictions by raising the baseline of trust and interoperability ahead of the next operation."[7] Contributions to multinational operations have become increasingly diverse, but a number of nations have extensive experience of working together. Analysts of strategic culture have as a result recognized increasingly that "ways of war," which have been viewed traditionally as being determined nationally, are ever more being shaped by multilateralism. Militaries, including air forces, have identified the need to train the way they fight. That logic can be extended to studies of military operations, requiring analysts to research the way that militaries fight, which is inherently multinational. The theoretical basis of interoperability is well understood, and although the literature on the conduct of multinational operations is increasing, there is scope for further research into the mechanics of those endeavors.

Many histories are focused on national perspectives for justifiable and obvious reasons, but as Skyler Cranmer and Elizabeth Menninga have pointed out, "The experience coalition partners have with one another may be every bit as important as their individual experiences."[8] The prevalence and importance of multinational operations prompted Olivier Schmitt to advocate the need for a "better understanding of their hierarchical dynamics, and the practicalities of military integration in a changing technological environment."[9] Although multinational operations have become commonplace, few detailed studies have been undertaken of international cooperation during air operations over a prolonged period. Issues of multinational cooperation have been considered in a growing number of books, but most examine a single conflict.[10] Multinational cooperation during contemporary operations, in particular, is a topic that is increasingly being subjected to analysis.[11] Jean-Baptiste Jeangène Vilmer nevertheless felt compelled to lament in 2018: "Too few studies concern themselves with how alliances and coalitions actually work in practice and how they perform when facing the challenges they were designed to address."[12]

This volume addresses case studies from around the globe, but it has a predominant focus on the experience of Western air forces, not least because of the frequency with which they have engaged in multinational operations. Stephen Cimbala and Peter Kent Forster have noted aptly: "From the beginning to the end of the Cold War and even afterward, burden sharing was the electric current that makes possible the congruity of aims and means by which Western democracies accomplished their stated purposes."[13] The United States, as the preeminent military power and technological leader, remains at the forefront of multinational operations and is the predominant focus of analyses of the topic. Arash Heydarian Pashakhanlou, for example, has called for "better burden sharing between the United States and its allies" to "improve the use of airpower in multinational coalitions."[14] European nations, perhaps with the exception of France, have in general "conceived their contributions in an American-led framework."[15] Nations that perceive themselves to have a "special relationship" with the United States—in particular one considered to be "an important key to their security and their political clout on the international scene"—have been "expected to be more willing to contribute meaningfully and shoulder greater risks in U.S.-led military interventions."[16] As a consequence, nations such as Australia and the United Kingdom are featured in a number of chapters in this collection, but the increasingly diverse nature of multinational operations is reflected in the later case studies.

This volume offers a comprehensive analysis of the dynamics of multinational air operations from the First World War to the contemporary era. The case studies traverse geographic regions and differing types of operations from major wars to peace operations. The analysis of United Nations (UN), North Atlantic Treaty Organization (NATO), and "coalition of the willing" operations allows for a comprehensive assessment of diverse multinational operations. Prior to the Second World War, interoperability was regarded largely as an "esoteric issue," but the significance of the concept was nevertheless recognized during the conflict.[17] The increasing frequency of multinational operations

US Air Force, US Marine Corps, and US Navy aircraft flying in formation with Japan Air Self-Defense Force and Royal Australian Air Force aircraft at Exercise Cope North in February 2018. Photograph by Senior Airman Jarrod Vickers, at https://www.pacom.mil/Media/News/News-Article-View/Article/1758417/cope-north-2019-strengthens-partnerships-sharpens-lethality-improves-interopera/.

prompted a sustained and intensifying drive to further interoperability between likely coalition partners. The trajectory has been far from straight as the introduction of new technology, the increasing diversity of coalitions, and the imposition of national caveats have created fresh challenges. There is a need, as a result, to study multinational air operations in both breadth and depth.

Bert Frandsen's chapter examines the transition of the United States from lacking a combat aviation arm on entry into the First World War to being involved in one of the largest multinational air operations of the conflict at St. Mihiel, France. Frandsen addresses the learning process as earlier British and French experience was applied as the scale, complexity, and integration of multinational air power increased, culminating in an air "armada" of more than 1,481 aircraft at St. Mihiel.

Although focused on the Battle of France during the Second World War, Matthew Powell's chapter provides important context on the prewar relationship between the British and French. Powell emphasizes that the absence of common doctrine and the lack of joint planning hindered operations in the face of the German onslaught, with the preexisting deficiencies being compounded by fraught command relationships. Andrew Conway then assesses the creation and development of the Anglo-American air component in the Western Desert between 1940 and 1942. Conway concludes that key logistical, operational, and political developments shaped the relationship between the Royal Air Force and the US Army Air Forces and that a combination of shared objectives and interpersonal relationships helped to ensure the fashioning and functioning of an effective military coalition. John Moremon rounds out the assessments of Allied air power in the Second World War by analyzing cooperation between the United States and New Zealand in the South Pacific Area. Cooperation during the Second World War pointed to the benefits of high-level cultural interoperability and strong personal relationships. Although the Royal New Zealand Air Force benefitted greatly from being reequipped with American aircraft, the operation of US Navy–type aircraft led to the New Zealanders being sidelined during the Bougainville campaign. As a rejoinder to the predominant focus on Allied air power in the scholarly literature, Stephen L. Renner addresses cooperation between Germany and Hungary during the Second World War, which resulted in the Royal Hungarian Air Force (Magyar Királyi Honvéd Légierő) becoming a de facto Luftwaffe auxiliary force. Issues of autonomy, command, and logistics were central to the relationship, which offers important insights into the challenges faced by a subordinate partner in alliance warfare.

Multinational cooperation continued into the Cold War, but some concepts and practices had to be relearned, as Corbin Williamson considers in his chapter on UN aircraft carrier operations during the Korean War. Although the Royal Navy, the US Marine Corps, and the US Navy were able to work together, attempts at de-confliction by separating areas of responsibility only limited rather than precluded the challenges caused by the use of different procedures and disparate training. Steven Paget's chapter on the integration of Number 2 Squadron, Royal Australian Air Force, with the US Air Force's 35th Tactical Fighter Wing during the Vietnam War highlights the advantages and drawbacks of multinational cooperation. A high degree of cooperation and successful operations masked a range of practical and conceptual challenges that resulted from the employment of nonstandardized aircraft, the utilization of differing rules of engagement, and logistical issues.

The Gulf War, as Richard P. Hallion surveys, was an inherently multinational endeavor in that approximately one-quarter of the coalition air strength (numerically) was provided by coalition partners of the United States. The much-heralded air campaign clouded some of the challenges encountered, including nonstandard equipment, doctrinal and training differences, dissimilar skill levels, divergent political mandates, and varying organizational cultures. The coalition's capacity to overcome these challenges in the main ensured that the campaign was still a success. In spite of the experience derived from the Gulf War, Maria E. Burczynska reflects that the highly successful NATO air campaign in Kosovo—Operation Allied Force—obscured a number of differences and capability gaps between the member states, which resulted in military efficiency being compromised in part to uphold alliance cohesion. Issues of command, disagreements over targeting, equipment deficiencies, and shortages affected the level of interoperability achieved. Benjamin S. Lambeth's chapter on Operation Iraqi Freedom emphasizes that the air campaign was a true coalition effort by Australia, the United Kingdom, and the United States. Early involvement in the planning of the campaign, previous experience, and efforts to generate interoperability were essential, but unanticipated problems were still encountered.

Finally, A. Walter Dorn examines UN air operations, which are inherently multinational. UN peace operations have been diverse in nature and have required the employment of all of the core roles of air power, including attack in some cases. Despite being underpinned by an organizational structure, these operations have been the subject of a range of challenges, including those of a logistical, financial, and ethical nature. Dorn demonstrates that although the UN has attempted to apply multinational air power in a judicious manner, its application has not been perfect.

Each chapter is important as a stand-alone case study in the conduct of multinational air operations throughout history. In addition, taken together, the collection of analyses provide a number of general principles about the challenges and opportunities presented by multinational operations. Lessons exist, but they must be contextualized. As Michael Howard explained, "The past is infinitely various, an inexhaustible storehouse of events from which we can prove anything or its contrary."[18] This volume does not constitute a handbook on how to conduct a multinational operation because the circumstances of each operation is unique. It does, however, outline the considerations that need to be taken into account when a multinational air operation is conducted.

A recent study on the primacy of air operations observed: "Improved targeting technology[, with in turn] the ability to apply force in limited ways[,] has increased the military effectiveness as well as the popularity of air power."[19] Air power is now the "'go-to' club in the decision-makers' golf bag,"[20] and it is delivered in a coalition context more often than not. An examination of the history of multinational air operations offers an opportunity to determine the variables that affect the level of interoperability achieved. Multinational operations are entrenched as part of the core businesses of air services, and this volume sheds light on the conduct and consequences of those endeavors.

Notes

1. Development, Concepts, and Doctrine Centre, *Joint Doctrine Publication 0-30: UK Air and Space Power* (Shrivenham: UK Ministry of Defence, 2017), 15.

2. Royal Australian Air Force, *Australian Air Publication 1000-D: The Air Power Manual*, 6th ed. (Canberra: Air Power Development Centre, 2013), 10.

3. Patricia A. Weitsman, *Waging War: Alliances, Coalitions, and Institutions of Interstate Violence* (Stanford, CA: Stanford University Press, 2014), 27.

4. Per Marius Frost-Nielsen, "Conditional Commitments: Why States Use Caveats to Reserve Their Efforts in Military Coalition Operations," *Contemporary Security Policy* 38, no. 3 (2017): 371.

5. Frank Ledwidge, *Aerial Warfare: The Battle for the Skies* (Oxford: Oxford University Press, 2018), 1.

6. Chiara Ruffa, "Military Cultures and Force Employment in Peace Operations," *Security Studies* 26, no. 3 (2017): 421.

7. Wing Commander André Adamson and Colonel Peter Goldfein, "The Trilateral Strategic Initiative—A Primer for Developing Future Air Power Cooperation," *Air Power Review* 19, no. 1 (Spring 2016): 81.

8. Skyler J. Cranmer and Elizabeth J. Menninga, "Coalition Quality and Multinational Dispute Outcomes," *International Interactions* 44, no. 2 (2018): 224.

9. Olivier Schmitt, "More Allies, Weaker Missions? How Junior Partners Contribute to Multinational Military Operations," *Contemporary Security Policy* (2018): 81, at https://www.tandfonline.com/doi/pdf/10.1080/13523260.2018.1501999?needAccess=true.

10. For example, see Jacob Neufeld and George Watson, eds., *Coalition Air Warfare in the Korean War, 1950–1953* (Washington, DC: US Air Force History and Museums Program, 2005); Karl Mueller, ed., *Precision and Purpose: Airpower in the Libyan Civil War* (Santa Monica, CA: RAND, 2015); Benjamin Lambeth, *NATO's Air War for Kosovo: A Strategic and Operational Assessment* (Santa Monica, CA: RAND, 2001); Benjamin Lambeth, *The Unseen War: Allied Air Power and the Takedown of Saddam Hussein* (Annapolis, MD: Naval Institute Press, 2013).

11. See John Andreas Olsen, ed., *Global Air Power* (Lincoln, NE: Potomac Books, 2011); John Andreas Olsen, ed., *European Air Power: Challenges and Opportunities* (Lincoln, NE: Potomac Books, 2014); A. Walter Dorn, ed., *Air Power in UN Operations: Wings for Peace* (Abingdon, UK: Routledge, 2016); Myron Hura, Gary W. McLeod, Eric V. Larson, James Schneider, Daniel Gonzales, Daniel M. Norton, Jody Jacobs, Kevin M. O'Connell,

William Little, Richard Mesic, and Lewis Jamison, *Interoperability: A Continuing Challenge in Coalition Air Operations* (Santa Monica, CA: Rand, 2000); Eric Larson, Gustav Lindstrom, Myron Hura, Ken Gardner, Jim Keffer, and William Little, *Interoperability of U.S. and NATO Allied Air Forces: Supporting Data and Case Studies* (Santa Monica, CA: RAND, 2003).

12. Jean-Baptiste Jeangène Vilmer, foreword to Olivier Schmitt, *Allies That Count: Junior Partners in Coalition Warfare* (Washington, DC: Georgetown University Press, 2018), x.

13. Stephen J. Cimbala and Peter Kent Forster, "The US, NATO, and Military Burden Sharing: Post–Cold War Accomplishments and Future Prospects," *Defense & Security Analysis* 33, no. 2 (2017): 117.

14. Arash Heydarian Pashakhanlou, "Air Power in Humanitarian Intervention: Kosovo and Libya in Comparative Perspective," *Defence Studies* 18, no. 1 (2018): 52.

15. Christian Anrig, "Air Power in Multinational Operations," in *Routledge Handbook of Air Power,* ed. John Andreas Olsen (Abingdon, UK: Routledge, 2018), 271.

16. Jens Ringsmose, "NATO Burden-Sharing Redux: Continuity and Change after the Cold War," *Contemporary Security Policy* 31, no. 2 (2010): 331; Stéfanie von Hlatky and Justin Massie, "Ideology, Ballots, and Alliances: Canadian Participation in Multinational Military Operations," *Contemporary Security Policy* 40, no. 1 (2019): 4.

17. Steven Paget, *The Dynamics of Coalition Naval Warfare: The Special Relationship at Sea* (Abingdon, UK: Routledge, 2017), 1.

18. Michael Howard, "The Lessons of History," *The History Teacher* 15, no. 4 (August 1982): 491.

19. Susan Hannah Allen and Carla Martinez Machain, "Choosing Air Strikes," *Journal of Global Security Studies* 3, no. 2 (2018): 150.

20. Clive Blount, "Useful for the Next Hundred Years? Maintaining the Future Utility of Airpower," *RUSI Journal,* June–July 2018, 2.

Building Partnership Capacity

US Air Power in the First World War

Bert Frandsen

Contemporary American military strategy emphasizes building partnership capacity. This concept involves advising, training, organizing, and equipping the military forces of an ally in need.[1] Examples include coalition assistance to build the air forces of Afghanistan and Iraq to help those countries defeat insurgencies and enable the exit of coalition forces. Something like this occurred during the First World War; only in this case the Americans were the recipients of military assistance. France, Britain, and Italy helped create American air power by training American pilots and equipping the American Expeditionary Forces (AEF) Air Service. France played the dominant role in this capacity-building effort. American air leaders learned how to conduct multinational air operations by serving under French Air Service command during the spring and summer of 1918. Americans subsequently exploited what they had learned and conducted one of the largest multinational air operations of the First World War at St. Mihiel, France.

What makes this case of capacity building so remarkable is that eighteen months before the US-led multinational air operation at St. Mihiel—when Congress declared war on 6 April 1917—the American air arm was nothing more than a small branch of the Signal Corps, and it was far behind the air forces of the warring European nations. The First World War, then in its third year, had witnessed the development of large air services with specialized aircraft for observation, bombardment, and pursuit. In contrast to the European air forces, an American combat aviation arm did not exist. When America entered the war, the Aviation Section of the Signal Corps consisted of twenty-six fully qualified aviators who were equipped with unarmed and, by European standards, obsolete airplanes. Some of the Aviation Section's pilots had, however, seen active service during the Mexican Punitive Expedition of 1916. The single squadron that accompanied this expedition deployed with eight unarmed, underpowered, and unreliable aircraft. Though unopposed in the air, the squadron proved incapable of effective mission performance.[2]

Within weeks of the US declaration of war in April 1917, French and British military officials arrived in Washington to urge the development of a large air force. A cable from French premier Alexandre Ribot was especially influential. He requested that by the spring of 1918 the United States have "4,500 airplanes, 5,000 pilots, and 50,000 mechanics" in France to help the Allies gain air supremacy.[3] The French request stimulated a "record-breaking" $640 million appropriation by the US Congress to build a mighty air force. French Army representatives even helped Signal Corps officials craft the legislation.[4] A wildly optimistic mobilization plan called for 263 combat squadrons in Europe by June 1918, but doing so would prove impossible given America's poor aviation infrastructure.[5] The building of an American air arm capable of combat operations in 1918 obviously required Allied assistance.

Equipping and Training the US Air Arm

Signal Corps officials sent a contingent of some 100 officers, technicians, and experts under Major Raynal Bolling to Europe to survey the aviation situation and determine what types of airplanes US factories should manufacture. Bolling was an excellent choice for the job because of his expertise in both legal and aviation matters. He had organized the first aviation company in the New York National Guard while also serving as general counsel for US Steel, the largest corporation in the world. After meeting with aviation officials in Britain, France, and Italy, Bolling realized that US aviation technology was so far behind the current standards that it would be necessary, at least initially, to rely upon the Europeans for airplanes.[6] At this point in aviation history, the airplane reflected an immature technology, and, unlike the case today, significant improvements were inexpensive and frequent. Moreover, the proximity of European aircraft designers and factories to the fighting provided a distinct advantage in turning out improved models based on front-line experience. Indeed, American industry had so much difficulty producing acceptable

Table 2.1. Sources of Front-Line Aircraft for the AEF Air Service in France

Source	Number of Aircraft	Representative Types
France	4,879	Nieuport 28, SPAD XIII, Breguet 14, Salmson 2A2
Britain	283	Sopwith Camel, SE-5
Italy	19	Caproni Bomber
United States	1,443	DH-4

Source: Numbers of aircraft gathered from the chart "Airplanes Received from all Sources, from Beginning of Operations to Dec . 31. 1918," in *The US Air Service in World War I,* vol. 1, ed. Mauer Mauer (Washington, DC: Office of Air Force History, 1978), 66. Representative types of aircraft were gathered from "Total Number of Airplanes Dispatched to Zone of Advance," in *The US Air Service in World War I,* ed. Mauer, 1:25.

warplanes that most of the AEF's airplanes came from foreign sources. The contract Bolling negotiated with the French called for 875 training planes and 5,000 service-type aircraft. French manufacturers were unable to deliver on time, however, resulting in some aircraft purchases by the United States from Britain and Italy (see table 2.1). France supplied some 80 percent of the AEF's airplanes.[7]

Pilot training also required foreign assistance. Only the most basic pilot training could take place in the United States because of the lack of suitable aircraft and instructors. As a consequence, Signal Corps officials developed a bifurcated training system in which preliminary training took place in the United States and advanced training overseas, though this was not always the case, as explained later in this chapter.[8] The proximity of a Royal Flying Corps (RFC) training program in Canada and the advantage of a common language help explain why British influence dominated the expansion of pilot training in the United States.

A month after the US declaration of war, Signal Corps representatives visited an RFC training facility in Toronto, Canada. They planned to implement a new ground phase of training modeled after the RFC's School of Military Aeronautics at the University of Toronto. The new ground phase would help weed out those personnel unsuitable for flight training (the Signal Corps had previously selected pilot trainees from serving officer volunteers). The new phase would also provide basic military training and education on the mechanics of flight. The US delegation included representatives from six

American colleges that had strong engineering programs. This visit led to the establishment of aviation cadet ground schools at eight American colleges (the Universities of California, Texas, Illinois; Ohio State; Cornell; Massachusetts Institute of Technology; and Georgia School of Technology). The Signal Corps provided these institutions with copies of the RFC's lectures to build their curriculum.[9]

After completing ground school, cadets progressed to preliminary flight training. The Signal Corps expanded the number of training sites in the United States from four to some thirty by the end of the war, but an initial lack of instructor pilots and training aircraft delayed large-scale pilot production.[10] The proximity of the RFC's training facilities in Canada again provided the opportunity to expand pilot training. A "Reciprocal Training Agreement" called for the RFC to train ten American squadrons in return for the RFC's use of airfields near Fort Worth, Texas, where better weather enabled year-round training. Three hundred American aviators received training in the Reciprocal Training Program.[11] A side benefit of the agreement was the transfer of several experienced RFC officers to leadership positions in the AEF Air Service. Canadian Harold Hartney transferred from the RFC to command the 27th Aero Squadron and would rise to command one of the three pursuit groups in the AEF before St. Mihiel.[12]

The lack of pilot instructors and training sites also spurred agreements in which the United States sent cadets overseas for basic flight training. Approximately 500 Americans trained in Great Britain. Of this total, 96 joined the AEF in France, and 216 served in the Royal Air Force (RAF), including two American squadrons, which were trained and equipped by the British and attached to the RAF in France until near the end of the war.[13]

The bulk of AEF advanced pilot training took place in France and, until the AEF could generate experienced instructors, under French instructors who usually could not speak English. Although 94 percent of American senior officers reported that language differences posed either no difficulty or only minor difficulties, the language barrier could be troublesome for American student pilots who could not speak French.[14] French instructors used hand signals to provide rudimentary instructions. In some cases, misunderstandings led to a student's failure.[15] Language difficulties were mitigated, however, by the French training system, which did not use dual-control aircraft to train pursuit pilots. Students instead progressed from taxiing airplanes that could not fly to successively piloting more advanced aircraft and to performing more advanced skills, such as aerobatics, without an instructor accompanying them in the air. The AEF established aviation instruction centers for observation, bombardment, and pursuit-pilot training in France. The largest center, at Issoudun, trained more than 1,000 pilots by midsummer of 1918.[16]

Italy also trained American pilots. The Italian Army constructed a training center at Foggia and provided experienced flight instructors. A total of 407 US pilots graduated from the school at Foggia, including 121 who completed bombardment training. Ninety-six of them served with Italian squadrons.[17] By the end of the war, the AEF Air Service operated ten schools in Europe, seven of which were constructed by American forces.[18]

ORGANIZING FOR COMBAT IN FRANCE

One of the challenges in forming combat-capable aero squadrons was finding experienced leaders. Although several RFC officers transferred to the AEF Air Service, a larger pool of talent existed among American volunteers flying with the French. The French Army established a squadron of American volunteer pilots in 1916. The exploits of this highly publicized pursuit squadron, called the Lafayette Escadrille, attracted even more American volunteers to fly for France. Like the pilots of the Lafayette Escadrille, most of these volunteers also served in pursuit squadrons but were scattered among many different French squadrons. Ninety-three of these pilots joined the AEF, distributing their experience throughout its air service.[19]

Capacity building was required for the entire AEF, not just for its air service, because the United States was not prepared for war. As a consequence, General John J. Pershing, the commander of the AEF, decided to conduct the final organization, training, and equipping of his units in France. Because the AEF operated within the French Army zone of operations, he also adopted French organizational structures that still exist in the US military, such as the numerical designation for staff organizations (1-personnel, 2-intelligence, 3-operations, etc.).[20] This is also why the US Air Force's organizational hierarchy ascends from squadron to group to wing, unlike the British system, which goes from squadron to wing to group.[21] Pershing decided to organize his air service as a separate combat arm, similar to the Allies, and detached it from the Signal Corps upon a recommendation from Major William "Billy" Mitchell.[22]

Mitchell was one of the first members of the Aviation Section to arrive in France. He would finish the war as a brigadier general and the AEF's senior operational air commander. He could speak French and was initially posted to France as an aviation observer. He toured the front, took detailed notes, and learned about air strategy, tactics, and organization through repeated visits with the French and British aviation commanders.[23] He was especially influenced by British general Hugh Trenchard's emphasis on the offensive use of air power.[24] By the summer of 1917, Colonel Mitchell served as Chief of Air Service, Zone of Advance. His responsibilities included the creation of organization and training centers where combat-ready aviation units were formed. The AEF established organization and training centers for pursuit, observation, bombardment, and balloon units. Trained pilots, mechanics, airplanes, and equipment came together at these centers and conducted final training before deploying to the front.[25]

Mitchell's headquarters consisted of twenty-four American officers, one British liaison officer, and a French contingent of eight officers. The French contingent under Major Paul Armengaud, the chief of the French Military Mission to the AEF Air Service, was the most influential.[26] Mitchell met Armengaud, who was serving as aeronautical officer with the French Army Group headquarters, shortly after arriving in France and judged him a man of exceptional ability. So close was their relationship that when the French Army tried to appoint a colonel to lead the advisory mission, Mitchell prevailed in retaining Major Armengaud as the senior adviser. The French Army also provided Mitchell with an interpreter, about whom Mitchell noted, "It was largely due to him that my staff never had a misunderstanding with the Allies all during war."[27]

The 1st Pursuit Organization and Training Center exemplifies how the French Air Service assisted in preparing units for front-line duty. Mitchell arranged for the center's establishment at the airfield used by Escadre (Wing) 1, France's largest fighter unit, to facilitate the training and equipping of the AEF's first pursuit units.[28] Major Bert Atkinson, one of the aviators on the Mexican Punitive Expedition, oversaw the construction of barracks and hangars at Villeneuve-les-Vertus. On 9 February 1918, Major Armengaud inspected the 1st Pursuit Organization and Training Center and declared it ready to begin forming pursuit squadrons. He informed Atkinson that he would soon receive trained pilots and Nieuport 28 airplanes. The fact that a French adviser, not an American officer, conducted the inspection of 1st Pursuit Organization and Training Center and set events in motion for the formation of America's first pursuit squadrons illustrates how closely the French Air Service oversaw the development of the AEF's pursuit capability.[29] Atkinson submitted his plan for training and organizing pursuit units to the Zone of Advance headquarters, where Armengaud's fighter expert and a former US Lafayette Escadrille pilot reviewed and modified it.[30]

The training program included joint patrols with Escadre 1's experienced pilots. One or two of Escadre 1's SPAD 7s escorted two or three Nieuport 28s on a patrol of the front.[31] These joint patrols ceased when the Germans launched their much-anticipated Spring Offensive on 21 March, and Escadre 1 left Villeneuve to operate against the German offensive near Amiens. Atkinson was fortunately able to arrange for several former Lafayette flyers who had transferred into the AEF to assume key leadership positions in the first squadron he would deploy to the front.[32]

GAINING EXPERIENCE UNDER FRENCH COMMAND

Germany's Spring Offensive of 1918 took advantage of the recent defeat of Russia on the eastern front to concentrate against the British and French in the west in hopes of winning the war before US entry could tip the scale in favor of the Allies. Using new stormtroop tactics, the Germans achieved great success. The resulting emergency for the Allies led to the appointment of French general Ferdinand Foch as supreme commander and to the occupation of trenches in quiet sectors by newly trained US divisions in order to release French divisions to counter the German advances. When the US 26th Division occupied trenches in the quiet Toul sector, observation and pursuit units were deployed from their organization and training centers to provide support, heralding the first employment of US air and ground units together at war. AEF units in this sector came under the operational control of the French 8th Army and the administrative control of the US 1st Corps.[33]

German fighter opposition in the Toul sector was weak and not equipped with the latest aircraft because the German air force had concentrated near Amiens for its offensive. The enemy air situation was just challenging enough for the inexperienced American aviation units to gain combat experience while having a good probability of survival. The 1st Pursuit Group claimed twenty-seven confirmed victories in the Toul sector and lost nine pilots, five of them as the result of accidents. Most of these victories were won by new pilots, not by the ones with experience flying for the French or British, thus attesting to the validity of the training program. The AEF Air Service produced its first aces, Eddie Rickenbacker and Douglas Campbell, at Toul.[34]

With the 1st Pursuit Group dominating the skies in the Toul sector, the 1st Corps Observation Group's three squadrons gained little experience operating in a contested airspace. They did, however, practice reconnaissance, artillery adjustment, and infantry-liaison techniques using French doctrinal procedures that had been translated into English. Most of the observers (observation aircraft had a two-man crew, observer and pilot) had also spent one or two months with French squadrons at the front. Unlike the pursuit group, the commanders and pilots of these three observation squadrons, with a few exceptions, lacked combat experience.[35] As historian James J. Cooke notes, "French flying officers were in evidence and, as at Issoudun, did a great deal of advising and actual training of the aero squadrons."[36]

One of the 1st Corps Observation Group's three squadrons, the 12th Aero Squadron, was sent to the Baccarat sector, about 33 miles (53 kilometers) east of Toul, to support the US 42nd Division, which had also recently occupied the trenches. In this case, the US squadron operated under the operational control of the French 6th Corps Chief of Air Service. The 12th Aero flew missions similar to the ones in the Toul sector: routine reconnaissance and artillery adjustment in support of the 42nd Division and a few corps-directed reconnaissance missions. The squadron reported that it received "much valuable advice and aid" from the French corps Chief of Air Service and the veteran observer he placed at the disposal of the squadron commander.[37]

In addition to the proficiency that aircrews gained in the Toul sector, the group and squadron commanders experienced the processes and products used by the French Air Service to plan, coordinate, and execute unit-level air operations. The French 8th Army's air service orders, air intelligence bulletins, and general intelligence summaries provided a model for the Americans to emulate. They also learned the importance of liaison with the front-line units, who could provide maps and diagrams to increase their situational awareness.[38] The most senior student of these French-directed air operations was Mitchell, recently appointed Chief of Air Service of the US 1st Corps. The Toul experience provided him with the opportunity to observe multinational air operations, begin to formulate his own command processes, and gain the measure of his men and their aircraft.

By the end of May 1918, the Germans had driven a deep wedge into the Marne sector, threatening Paris. The emergency caused Pershing to agree to commit the US 2nd and 3rd Divisions to reinforce the French 6th Army to help stop the German advance at Château-Thierry. The Marine Brigade of the US

2nd Division took particularly heavy casualties in one of its most famous battles, Belleau Wood.[39] Colonel Walter Grant, Pershing's deputy chief of staff, observed the effect of German air superiority at Belleau Wood and sent a message back to AEF headquarters recommending commitment of air units to the battle.[40] That same day, French air advisers Major Armengaud and Captain Dudoré (full name not known) appeared at the US 1st Army Air Service headquarters in Toul with instructions to send aviation units to the Marne sector.[41]

Mitchell took charge of the newly created 1st Air Brigade sent to reinforce the French in the Marne sector.[42] The four squadrons of the 1st Pursuit Group and two squadrons of the 1st Observation Group arrived at airfields behind Château-Thierry at the end of June. A French observation squadron, already on station, rounded out the 1st Observation Group. On 1 July, these units supported the US 2nd Division's attack on Vaux. Unlike the action at Belleau Wood, the division enjoyed the advantages of air superiority. The 1st Pursuit Group prevented hostile aircraft from attacking American ground troops and enabled observation planes and balloons to adjust artillery, conduct reconnaissance, and report on the progress of the attack.[43] Mitchell's brigade performed brilliantly. In a matter of days, it had moved some 130 miles (209 kilometers) to new airfields, established a new line of logistical support, and fought America's first air–land battle. Pershing later called the battle at Vaux a "brilliantly executed operation."[44]

The complexity of command relationships during these multinational air operations confused Mitchell's subordinates. The US 1st Pursuit Group replaced the French 6th Army's Pursuit Group and received orders from the 6th Army's Chief of Air Service. However, the Chief of Air Service of the US 1st Corps also expected the pursuit group to follow the 1st Corps' orders because it, commanded by a general officer, was the senior US headquarters in the area. Adding to the confusion were orders to the group from the ever-dynamic Colonel Mitchell as brigade commander. According to the operations officer of the 1st Pursuit Group, "The suggestions of the Brigade Commander were often a source of extreme worry" for the leaders of the 1st Pursuit Group, who found it "impossible to obey both the orders of the 6th French Army and the suggestions of the Commanding Officer of the First Brigade. . . . [W]hether the Brigade's functions were tactical or administrative was never made clear."[45] To be sure, the US Army was still working out the nuances of command relationships between the pursuit and observation groups and among the armies and corps they supported, and this process was further complicated while under French Army command.

French-Led Multinational Air Operation on the Marne

Anticipating a renewal of the German offensive in the Marne sector, Allied commander in chief General Ferdinand Foch reinforced five French armies with nine US divisions and two British divisions for a two-phased battle that would determine the future course of the war. The first phase required defeating a double envelopment of Reims by three German armies. The second phase was to counterattack and crush the Germans in the Marne salient.[46] Foch also concentrated multinational air forces, making the Second Battle of the Marne the largest air battle of the war, according to a study by E. R. Hooton. Allied air forces included the French Air Division, the Franco-Italian heavy-bomber force, the RAF's 9th Brigade, and the US 1st Pursuit Group as well as the organic aviation units of five French armies and their corps. The French Air Division, created on 14 May 1918 and commanded by General Maurice Duval, was the largest aviation unit of the war. Its two brigades consisted of some 370 fighters and 230 bombers. The RAF's 9th Brigade provided an additional nine squadrons (five fighter, four bomber) of offensive air power. In addition, Mitchell's 1st Air Brigade and the organic aviation units of five French armies made for a total of some 1,800 Allied and 900 German airplanes in the Second Battle of the Marne.[47]

Duval's orders called for fighters to seize air superiority while bombers attacked German bridging efforts across the Marne.[48] A brilliant defense in depth by the French 4th Army stopped the German offensive east of Reims. West of Reims, the German 7th Army succeeded in pushing across the Marne but was subject to a gauntlet of artillery fire and air attacks. Multinational air forces flew 800 sorties during the first day, dropping nearly fifty-three tons of bombs in low-level attacks on the bridges. A captured German message referred to the bombardment as a veritable hell. The chief engineer of the German 7th Army was killed as he observed bridge construction over the river. Seventy percent of the German bridges were destroyed after two days, and the German offensive was defeated.[49]

The Allies did not conduct the largest multinational air operation of the war up to that point without Clausewitzian friction. In an unusual turn of events, the US 1st Pursuit Group received no orders for the German attack on 15 July, in spite of accurate intelligence warning of an imminent attack. As a consequence, the group commander kept three squadrons on alert (a fourth was being reequipped with the SPAD 13) as he awaited orders. Late in the

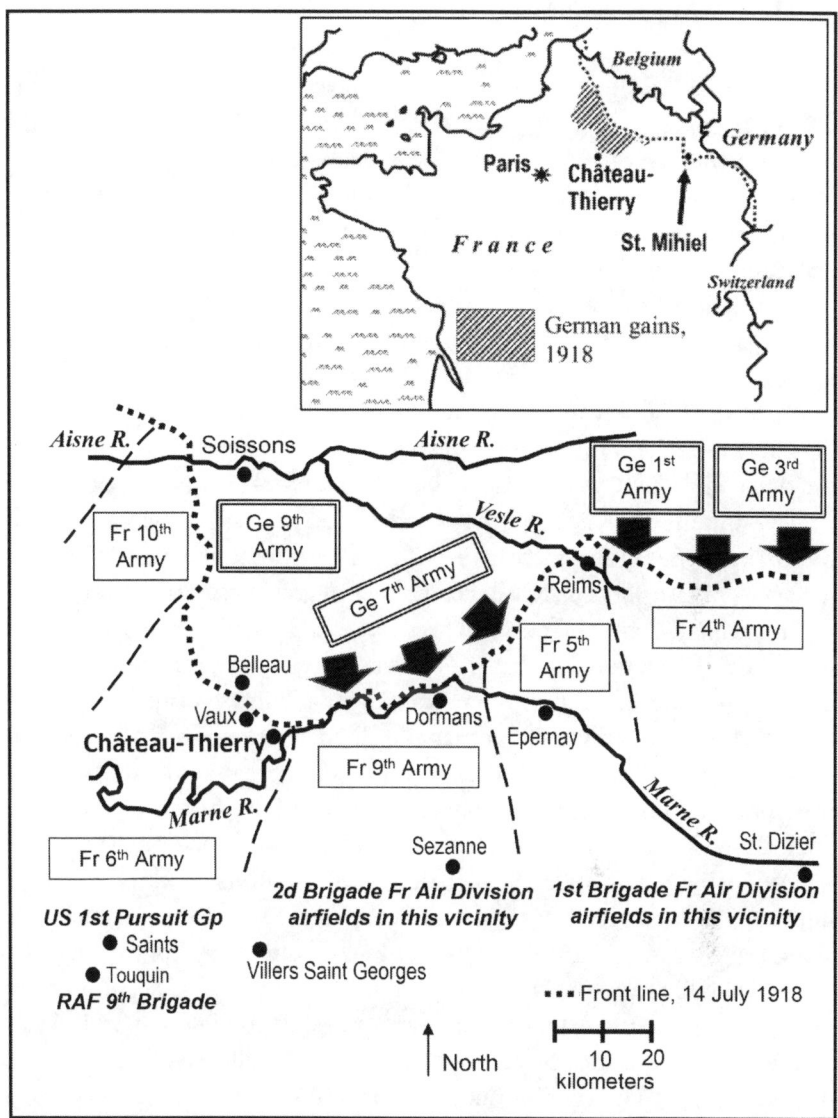

Second Battle of the Marne. The French Army massed a multinational air strike force for this battle. The French Air Division, the RAF 9th Brigade, and the US 1st Pursuit Group deployed to these locations before the battle. Map created by Bert Frandsen based on data from "Récapitulation schématique des opérations du 18 juillet au 25 septembre 1918," map 32 in État major de l'armée, Service historique, *Cartes,* vol. 1 of *Les Armées françaises dans la Grande Guerre* (Paris: Imprimerie nationale, 1938), and map 71 in *Military Operations of the American Expeditionary Forces,* vol. 5 of *The United States Army in the World War, 1917–1919* (Washington, DC: Center of Military History, US Army, 1989), 231.

day, the group launched two squadron-size missions into the battle area. One US pursuit squadron escorted an attack by RAF 9th Brigade bombers on the bridges, while the other squadron flew air superiority patrols.[50] The reason for such a meager and late effort by the Americans on the first day of an important battle was, according to Mitchell, that French Army aviation officials had yet to complete the preparations required to coordinate the efforts of three national aviation contingents.[51] Perhaps the late addition of the RAF 9th Brigade, which arrived at its airfields near Touquin only on 14 July after flying through rainstorms, contributed to the delay.

As a result of previous combined operations with the RAF, French planners were mindful of the complexities and potential dangers inherent in multinational air operations. Fratricide had been a problem when the RAF 9th Brigade and the French Air Division cooperated against a previous phase of the German offensive in early June in the Noyon-Montdidier sector. In a few instances, British pilots had mistakenly attacked French troops on the ground, and French pilots and infantry had fired upon RAF planes. Such occurrences impressed commanders with the necessity for close coordination in combined air operations. As a further precaution, the RAF 9th Brigade also instituted the practice of flying its airplanes to French airfields before combined operations to give the Allies an opportunity to familiarize themselves with the type of aircraft the brigade's pilots flew. On 15 July, the 9th Brigade conducted such a demonstration over the 1st Pursuit Group's field at Saints.[52]

While the British stunted over Saints, French officers developed plans to integrate all of the Allied aviation contingents into a massive aerial operation along the Marne. Orders issued by the French 6th Army coordinated the efforts of the French Air Division, the British 9th Air Brigade, and the 1st Pursuit Group in supporting a counterattack against the German bridgehead on the south side of the Marne. Sixth Army planners allocated specific times in which each national contingent would be responsible for aerial operations in the battle area and outlined the tasks to be accomplished during the allotted period. The length of time allocated to each contingent depended on its relative size: the larger the national contingent, the more time was allocated for its operations. The 1st Pursuit Group assisted the multinational effort over the Marne bridgehead by flying air superiority missions to protect friendly observation aircraft and balloons. It destroyed six enemy aircraft and one balloon and conducted some strafing against ground targets on 16 July. At the end of the day, nine of the group's pilots were missing. Fortunately, six of them turned out to be missing because of forced landings due to damaged aircraft or wounded pilots.[53]

Foch's counteroffensive from both sides of the salient on 18 July struck just as the German offensive culminated. The French 10th and 6th Armies, constituting the main effort, attacked from the previously quiet west side of the salient. A brigade of the French Air Division supported the 10th Army, while the RAF 9th Brigade and 1st Pursuit Group supported the 6th Army. The Allied counterstroke achieved surprise and enjoyed the advantages of air superiority and a powerful tank force against no German tanks.[54] Although the Germans' resistance stiffened, by the end of the month they withdrew from the salient. The initiative passed to the Allies, and the Germans never regained it. The French Air Division and RAF 9th Brigade subsequently left the Marne sector to support the British Army's August counteroffensive toward Amiens.[55]

Mitchell's presence at the Marne enabled him to organize a tactical head-quarters and observe how French aviation officials commanded multinational air operations during high-intensity combat. He regularly visited the French 6th Army Air Service headquarters as well as Foch's supreme head-quarters, where Mitchell's French aviation adviser, Major Armengaud, was posted during this critical battle. Such visits helped Mitchell develop a strategic appreciation of operations and further strengthened his relationship with Armengaud.[56] The transfer of AEF aviation officers who fought in the Second Battle of the Marne to leadership positions in new aviation units being organized in the Toul sector extended experience in multinational air operations to the rest of the AEF Air Service. This transmission of experience is best exemplified by the transfer of the commander and operations officer of the 1st Pursuit Group to command of the 1st Pursuit Wing, which consisted of two new pursuit groups and a bombardment group (a fighter-bomber organization fashioned after French and British formations taking part in the Marne campaign).[57] The knowledge gained by American Air Service leaders in multinational air operations on the Marne would be put to good to use at St. Mihiel, where an American field army made its debut.

THE US-LED MULTINATIONAL AIR OPERATION AT ST. MIHIEL

Pershing had long planned for a major offensive against the St. Mihiel salient in the Toul sector, where AEF units had been training. The salient measured 15 miles (24 kilometers) deep and 25 miles (40 kilometers) wide. It was a strong defensive position that the Germans had improved for years. Seven German divisions defended the salient with support from approximately 150 pursuit, 120 reconnaissance, and 25 battle planes. Foch wanted Pershing to finish this battle quickly and limited the objective to merely eliminating the salient because he wanted Pershing's main offensive to take place in the Meuse–Argonne sector. To help Pershing finish quickly, the French reinforced the US 1st Army with troops and combat multipliers, including tanks, artillery, and aviation. Pershing's plan called for a pincer attack, with the US V Corps (including one French division) attacking from the west and the US I and IV Corps attacking from the south. The French II Corps in the center would make holding attacks to keep the Germans from withdrawing.[58] French crews manned one-third of the artillery pieces, 113 of 267 tanks, and most of the aviation.[59]

On 20 August, Mitchell, now Chief of Air Service of the US 1st Army, learned from Armengaud that the French Army would provide aviation reinforcements.[60] The French Air Division and the Franco-Italian Night Bombardment Group added significantly to Mitchell's striking power. The French also provided four observation squadrons, enabling Mitchell to attach at least one experienced French observation squadron to each of his three corps observation groups, addressing one of his chief concerns—the lack of combat experience in his recently organized units. The French II Colonial Corps also came with its own aviation support, consisting of four observation squadrons and three balloon companies. Another four French squadrons formed an artillery adjustment group to regulate army-level artillery, including long-range guns capable of reaching the Metz fortifications behind St. Mihiel.[61] General Hugh Trenchard, commanding the RAF Independent Force, contributed another eight bombardment squadrons in support, though not under Mitchell's command.[62] The total force numbered 1,481 aircraft and twenty-one balloon companies (fifteen American, six French), a huge numerical advantage over the Germans. The AEF Air Service constituted about 40 percent of this multinational force.[63]

The commanders of the RAF Independent Force and French Air Division conducted coordination visits with the US 1st Army in the days leading up to the St. Mihiel offensive. French advisers also continued to have a strong influence on the AEF Air Service. Mitchell appointed Armengaud as his assistant chief of staff and kept him close, sharing his office with the adviser. Mitchell also appointed another member of the French mission, Captain K. Flury,

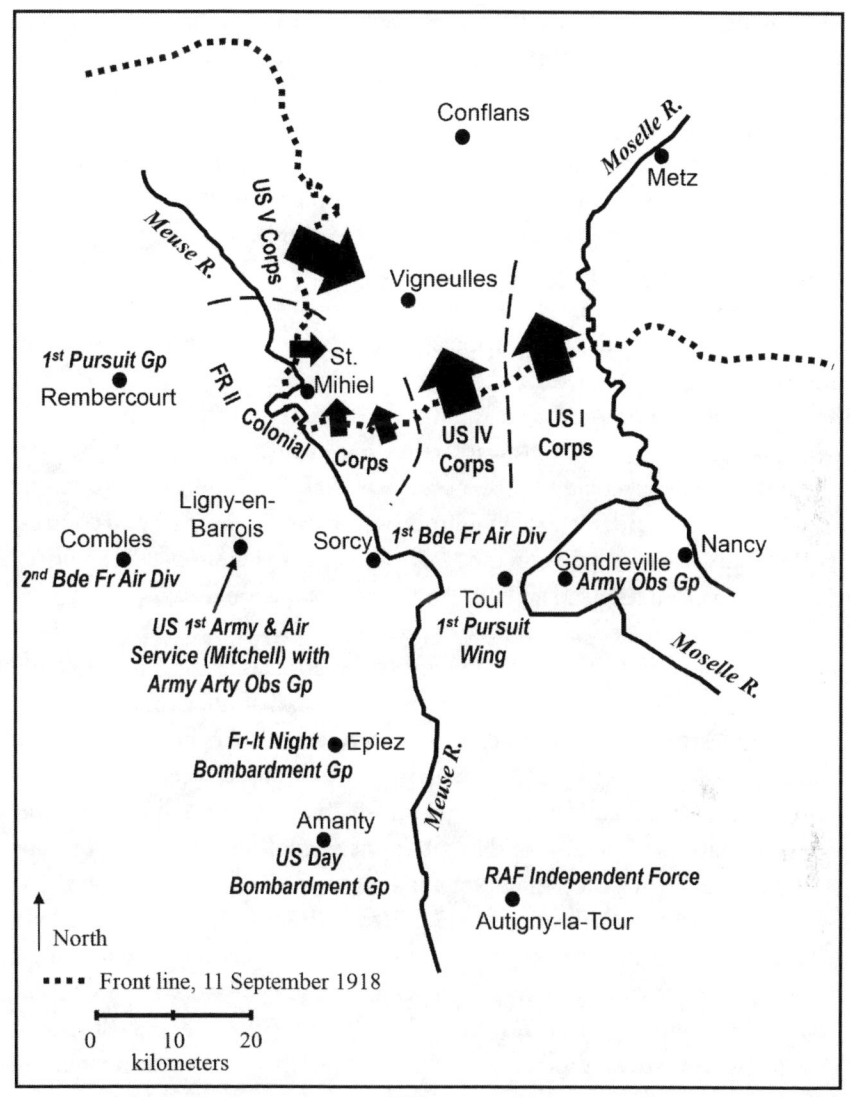

Conflans

Moselle R.

Metz

US V Corps

Meuse R.

Vigneulles

1st Pursuit Gp

Rembercourt

FR II Colonial Corps

St. Mihiel

US IV Corps

US I Corps

Ligny-en-Barrois

Combles

2nd Bde Fr Air Div

Sorcy

1st Bde Fr Air Div

Gondreville

Army Obs Gp

Nancy

US 1st Army & Air
Service (Mitchell) with
Army Arty Obs Gp

Toul

1st Pursuit
Wing

Moselle R.

Fr-It Night
Bombardment Gp

Epiez

Meuse R.

Amanty

US Day
Bombardment Gp

RAF Independent Force

Autigny-la-Tour

North

•••• Front line, 11 September 1918

0   10   20
kilometers

Battle of St. Mihiel. French, British, and Italian aviation forces reinforced the US 1st Army. Map displays locations of major headquarters supporting Colonel William Mitchell's multinational air offensive. Map created by Bert Frandsen based on data from William Mitchell, "The Air Service at St. Mihiel," *World's Work* 38 (August 1919): 365, and "Combined Order of Battle, St. Mihiel Operation," map 110 in *Military Operations of the American Expeditionary Forces*, vol. 8 of *The United States Army in the World War, 1917–1919* (Washington, DC: Center of Military History, US Army, 1990), 279.

as an assistant operations officer. The French Air Division posted Captain Jacques de Sieyes, who had spent 1917 at Ft. Sill in the United States teaching observation students, as its liaison officer.[64] Coordination measures included tying the Air Division into the 1st Army wireless net, providing instructions on reporting requirements, scheduling meeting times for liaison officers, and supplying targeting information to Allied bombing units.[65]

Mitchell and Armengaud conducted a reconnaissance flight over the salient on 10 September and reported to Pershing that the Germans appeared to be beginning to withdraw. Indeed, the Germans knew of the impending attack and would conduct a delaying action. The next day Mitchell assembled officers from each of the organizations under his direction—British, French, Italian, and American—and read them the operation order that he had written personally. A twelve-by-twelve-foot relief map courtesy of French balloon units aided his presentation. To ensure everyone understood the order, he asked each commander to explain how he would execute his part.[66] In contrast to the Allied defensive battle on the Marne, Mitchell's plan supported an offensive and therefore reflected an entirely different operational approach.

Mitchell's concept called for his air forces to directly support the army's advance while also fighting a deep battle to interdict, disrupt, and force the German air force to defend its own rear area instead of interfering with Pershing's offensive. His order stated, "Our air service will take the offensive at all points with the object of destroying the enemy's air service, attacking his troops on the ground and protecting our own air and ground troops." American reconnaissance, pursuit, and bombardment aviation would directly support the ground attack. Mitchell's veteran 1st Pursuit Group, now commanded by Canadian Harold Hartney, operated from the west side of the salient. Mitchell's new 1st Pursuit Wing, consisting of two American pursuit groups, a French pursuit group, and the US Day Bombardment Group, operated from Toul and surrounding airfields against the south side of the salient. American pursuit planes would protect friendly observation aircraft and troops by maintaining a "barrage" at multiple altitudes 2 to 3.5 miles (3.2 to 5.6 kilometers) in advance of the attack, while also destroying enemy balloons and attacking retreating enemy columns. One of the 1st Pursuit Wing's three pursuit groups was armed with bombs and held in reserve to conduct ground attacks on Mitchell's orders.[67]

Mitchell's coalition reinforcements would conduct deep attacks against the back of the salient extending another 15 to 20 miles (24 to 32 kilometers) into the German rear. The Air Division would conduct brigade-size attacks along each side of the salient against enemy aircraft and ground troops. These attacks were to be timed like roundhouse punches—as one roundhouse punch was finishing from one side of the salient, another would commence from the opposite side to catch enemy fighters in the rear.[68] Day and night bombardment aircraft of the Air Division, the Franco-Italian Bombardment Group, and Trenchard's Independent Force would strike railroad centers, airfields, ammunition dumps, and troop concentrations reaching back toward Metz.[69]

In spite of bad weather, Mitchell's plan worked well. His numerically superior force seized the initiative, forcing the air battle to occur well to the German rear. His aerial offensive enabled observation and pursuit aircraft to operate unhindered over the salient for the first two days of the attack. Multinational air attacks against German troops and supply points disrupted their withdrawal. Pershing's pincers advanced rapidly, eliminating the salient by the end of the second day and capturing 16,000 German soldiers.[70]

The Allied air forces flew 4,529 sorties during 12–16 September and lost 75 aircraft, a low loss rate for the western front. More than a third of Mitchell's losses were bombers—the US Day Bombardment Group suffered especially.[71] The operation also revealed significant shortcomings in US air–ground coordination, which would appear again in the Meuse-Argonne. Nevertheless, General Pershing complimented Mitchell on "the organization and control of the tremendous concentration" of the multinational air force that "enabled 1st Army to so successfully accomplish its dangerous and important mission."[72] The reduction of the salient secured important railway communications and restored 200 square miles (518 square kilometers) of French territory. More significantly, the victory demonstrated that a capable American field army had arrived on the Allies' side.[73]

By 15 September, Mitchell shifted his attention to making preparations for the Meuse-Argonne offensive, where the US 1st Army acted as the right wing of a general Allied offensive.[74] The French withdrew most of their aviation reinforcements to support their own offensive, leaving Mitchell with little more than half the force that he had at St. Mihiel to cover a larger sector and a much more determined German defense. Allies, primarily French, continued to provide some 200 of his 821 aircraft. His force's overall level of experience was also much reduced.[75] As a consequence, Mitchell was unable to replicate what he had achieved at St. Mihiel. The Meuse-Argonne was a difficult battle in the air and on the ground. A fuller accounting of this greatest American battle of World War I is beyond the scope of this chapter, but the Meuse-Argonne offensive succeeded in penetrating the Hindenburg Line and contributed to Allied victory.

Conclusion

In summary, the United States would not have been able to conduct air operations without the assistance provided by its French, British, and Italian allies in training and equipping the AEF Air Service. Key strategic decisions included the placement of the AEF within the French Army's zone of operations and Pershing's judgment to conduct the final organization and training of the AEF in France. These decisions enabled Allied, especially French, capacity-building efforts. The purchase of mainly French aircraft also facilitated interoperability as the United States fought its major operations with French Air Service reinforcements and French troops on each side.

The Allies were motivated to build American capacity because they needed it to help them win the war, and France was largely responsible for this effort. French advisory mission and liaison officers, though small in number, played a key role in the buildup. The association between Mitchell and Armengaud demonstrates the importance of personal relationships in such endeavors. The AEF's adoption of French Army organizational and staff structures and translated doctrines as well as its use of interpreters also facilitated multinational coordination. The experience American air leaders gained under French command at Toul and the Marne served as stepping stones to successful US-led multinational air operations in the final offensives of the war.

The air operation at St. Mihiel was particularly important because it demonstrated that a US field army capable of modern, combined arms combat had entered the war, lifting the Allies' morale while lowering the Germans'. Mitchell's orders presentation at St. Mihiel, along with requiring his subordinate and supporting commanders to explain how they intended to accomplish their respective missions, helped ensure a common understanding and achieve unity of effort. His concentration of multinational air power under a single commander would stimulate a vision for the centralized control of air power that would inspire his followers long after he departed the scene.

An evaluation of building partnership capacity in this case must acknowledge that the United States was ideally suited to be on the receiving end of

Allied capacity-building efforts. It was an advanced, industrialized nation with a well-developed military institution that could recruit its air officers from prestigious engineering colleges across the country. The small cadre of regular US Army air officers exploited their allies' assistance as they created a combat arm of the line and conducted multinational air operations. Under their leadership, the AEF Air Service grew to a force consisting of 6,861 officers, 51,229 men, forty-five squadrons, and twenty-three balloon companies in France, and the Air Service remaining in the United States was more than twice as large.[76] Considering the state of American military aviation when the United States entered the war, only eighteen months before St. Mihiel, this was indeed a remarkably successful case of building partnership capacity.

NOTES

1. Kathleen J. McInnis and Nathan J. Lucas, "What Is 'Building Partnership Capacity'? Issues for Congress," *Congressional Research Service,* December 18, 2015, 1.

2. Roger G. Miller, *A Preliminary to War: The 1st Aero Squadron and the Mexican Punitive Expedition of 1916* (Washington, DC: Office of Air Force History, 2003), 2, 51–53.

3. James J. Hudson, *Hostile Skies: A Combat History of the American Air Service in World War I* (Syracuse, NY: Syracuse University Press, 1968), 4.

4. Henry Greenleaf Pearson, *A Businessman in Uniform: Raynal Cawthorne Bolling* (New York: Duffield, 1923), 109.

5. Hudson, *Hostile Skies,* 6.

6. I. B. Holley, *Ideas and Weapons* (New Haven, CT: Yale University Press, 1953; reprint, Washington, DC: Air Force History and Museums Program, 1997), 53, 62.

7. *The Final Report of Chief of Air Service, AEF and a Tactical History of the Air Service, AEF* (originally published as *Air Service Information Circulars* in 1920 and 1921), vol. 1 of *The US Air Service in World War I,* 4 vols., ed. Mauer Mauer (Washington, DC: Office of Air Force History, 1978), 65–66; Pearson, *A Businessman in Uniform,* 175.

8. *Final Report of Chief of Air Service,* 1:93.

9. Hiram Bingham, *An Explorer in the Air Service* (New Haven, CT: Yale University Press, 1920), 14–31.

10. Rebecca Hancock Cameron, *Training to Fly: Military Flight Training, 1907–1945* (Washington, DC: Air Force History and Museums Program, 1999), 107, 121–23.

11. William E. Chajkowsky, *Royal Flying Corps: Borden to Texas to Beamsville* (Ontario: Boston Mills Press, 1979), 52; H. A. Jones, *The War in the Air: Being the Story of the Part Played in the Great War by the Royal Air Force,* vol. 5 (Oxford: Clarendon Press, 1935), 461–66.

12. Harold E. Hartney, *Up and at 'Em* (Harrisburg, PA: Stackpole, 1940), 15–18, 105–17.

13. *Final Report of Chief of Air Service,* 1:103.

14. Robert B. Bruce, *A Fraternity of Arms: America and France in the Great War* (Lawrence: University of Kansas Press, 2003), 149.

15. Charles Woolley, *First to the Front: The Aerial Adventures of 1st Lt. Waldo Heinrichs and the 95th Aero Squadron, 1917–1918* (Atglen, PA: Schiffer, 2004), 38.

16. Cameron, *Training to Fly,* 165–68.

17. Hudson, *Hostile Skies,* 34.

18. *Final Report of Chief of Air Service,* 1:112.

19. Roger G. Miller, *Like a Thunderbolt: The Lafayette Escadrille and the Advent of the American Pursuit in World War I* (Washington, DC: Air Force History and Museums Program, 2007), 1–2.

20. J. D. Hittle, *The Military Staff: Its History and Development* (Harrisburg, PA: Stackpole, 1961), 211.

21. Major William M. Mitchell, subject: Air Policy in France, 13 June 1917, Gorrell History of American Expeditionary Forces Air Service, 1917–19, series A, vol. 23, Record Group (RG) 120, National Archives and Records Administration (NARA), College Park, MD; Bert Frandsen, *Hat in the Ring: The Birth of American Air Power in the Great War* (Washington, DC: Smithsonian Books, 2003), 11–13.

22. William Mitchell, *Memoirs of World War I: From Start to Finish of Our Greatest War* (New York: Random House, 1960), 135.

23. Ibid., 21–30, 35, 48–53, 103–11.

24. Ibid., 111.

25. Lucien Thayer, *America's First Eagles: The Official History of the US Air Service, AEF 1917–1918*, ed. Donald J. McGee and Roger J. Bender (1919; reprint, San Jose, CA: R. James Bender, 1983), 95–100. AEF Air Service officials ordered Lieutenant Thayer to work on its official history effort in 1918.

26. James J. Cooke, *The US Air Service in the Great War, 1917–1919* (Westport, CT: Praeger, 1996), 77.

27. Mitchell, *Memoirs of World War I,* 99, 144.

28. Thayer, *America's First Eagles,* 115.

29. "Diary of First Pursuit Group," [AEF Air Service], 9 February 1918, Bert M. Atkinson Papers, Auburn University Library, Auburn, AL.

30. B. M. Atkinson to Assistant Chief of Air Service, Zone of Advance, memorandum, subject: Organizational Training for Pursuit Units, 6 April 1918, file 167.601–3, Frank P. Lahm Papers, USAF Historical Research Institute, Maxwell Air Force Base, AL.

31. Harold Buckley, *Squadron 95* (Paris: Obelisk Press, 1933), 33–34.

32. Memorandum listing pilots of 94th Aero Squadron and their schools and experience, 6 March 1918, Atkinson Papers; "Diary of First Pursuit Group," 7 March 1918; *Final Report of Chief of Air Service,* 1:285.

33. *Final Report of Chief of Air Service,* 1:29, 283.

34. Ibid., 1:138–43, 146–48.

35. Ibid., 1:171.

36. Cooke, *US Air Service in the Great War,* 61.

37. *Final Report of Chief of Air Service,* 1:190.

38. Ibid., 1:286.

39. Edward M. Coffman, *The War to End All Wars: The American Military Experience in World War I* (Madison: University of Wisconsin Press, 1986), 212–21; Bert Frandsen, "America's First Air–Land Battle," *Air and Space Power Journal,* Winter 2003, 32.

40. Colonel Walter S. Grant to [Colonel Fox] Connor, June 15, 1918, in *Early Military Operations of the American Expeditionary Forces*, vol. 4 of *The United States Army in the World War, 1917–1919*, 17 vols. (Washington, DC: US Army Center of Military History, 1988), 4:490.

41. Colonel Frank Lahm, *The World War I Diary of Col. Frank P. Lahm*, ed. Albert F. Simpson (Montgomery, AL: Historical Research Division, Maxwell Air Force Base, 1970), 90.

42. Mitchell, *Memoirs of World War I*, 208.

43. Elmer Haslett, *Luck on the Wing* (New York: Dutton, 1920), 77–78; Operations Report, Plans, and Training Officer, B Company, 23rd US Infantry, to Commanding Officer, 23rd US Infantry, in *Early Military Operations of the American Expeditionary Forces*, 4:665–66; Commanding General, 3rd Brigade, to Commanding General, 2nd Division, 2 July 1918, in *Early Military Operations of the American Expeditionary Forces*, 4:674–75.

44. John J. Pershing, *My Experiences in the World War* (New York: Stokes, 1931), 90.

45. Philip Roosevelt, "The Air Service in the Chateau Thierry Campaign, Part III, First Pursuit Group Administration," 1–2, Gorrell History, 1917–19, series C, vol. 1, Microfilm Publication M990, RG 120, NARA.

46. Bruce, *A Fraternity of Arms*, 226.

47. E. R. Hooton, *War over the Trenches: Air Power and the Western Front Campaigns 1916–1918* (Hersham, UK: Midland, 2010), 221, 228–35; Frandsen, *Hat in the Ring*, 179.

48. Hooton, *War over the Trenches*, 230.

49. Elizabeth Greenhalgh, *The French Army and the First World War* (Cambridge: Cambridge University Press, 2014), 308–9; David T. Zabecki, *The German 1918 Offensives: A Case Study in the Operational Level of War* (New York: Routledge, 2006), 262.

50. First Pursuit Group [AEF] Operations Report, 15 July 1918, Gorrell History, 1917–1919, series C, vol. 9, Microfilm Publication M990, RG 120, NARA.

51. Mitchell, *Memoirs of World War I*, 220.

52. H. A. Jones, *The War in the Air: Being the Story of the Part Played in the Great War by the Royal Air Force*, vol. 6 (London: Oxford University Press, 1937), 401–4, 412–13; "Order of Battle of the RAF on 8 August 1918," appendix 24 in H. A. Jones, *The War in the Air: Being the Story of the Part Played in the Great War by the Royal Air Force: Appendices* (London: Oxford University Press, 1937), 116–17; Hartney, *Up and at 'Em*, 162–63; Frandsen, *Hat in the Ring*, 182.

53. First Pursuit Group [AEF] Operations Order Number 37, 15 July 1918; First Pursuit Group Report of Operations, 16 July 1918; Report of First Pursuit Group for the Month of July 1918; First Pursuit Group Casualties, 12 December 1918, all in Gorrell History, series C, vol. 9, Microfilm Publication M990, RG 120, NARA. See also *U.S. Air Service Victory Credits*, USAF Historical Study no. 133, M-U 37218 (Montgomery, AL: Air University Library, Maxwell Air Force Base, 1969), 69–70; Hartney, *Up and at 'Em*, 168–69.

54. Zabecki, *The German 1918 Offensives*, 273.

55. Michael S. Neiberg, *The Second Battle of the Marne* (Bloomington: Indiana University Press, 2008), 177; Hooton, *War over the Trenches*, 236.

56. Mitchell, *Memoirs of World War I*, 211.

57. T. D. Milling, General Orders Number 1, HQs, Chief of Air Service, First Army, 28 August 1918, in *The Battle of St. Mihiel*, vol. 3 of *The US Air Service in World War I*,

ed. Mauer Mauer (Washington, DC: Office of Air Force History, 1979), 57; Philip Roosevelt, First Pursuit Wing Operations Memorandum No. 3, 29 August 1918, in *The Battle of St. Mihiel,* 3:68.

58. Mauer Mauer, editor's comments in *The Battle of St. Mihiel,* 3:v, 126, 137.

59. Bruce, *A Fraternity of Arms,* 262.

60. Lahm, *World War I Diary of Col. Frank Lahm,* 21 August 1918, 120–21; Commander in Chief, French Armies of the North and Northeast, Air Service, to Commander of the AEF GHQ, 24 August 1918, in *The Battle of St. Mihiel,* 3:54–55.

61. *Final Report of Chief of Air Service,* 1:37.

62. Hooton, *War over the Trenches,* 246.

63. *Final Report of Chief of Air Service,* 1:37, 383.

64. Mitchell, *Memoirs of World War I,* 237; T. D. Milling, General Order No. 1, Headquarters Chief of Air Service, First Army, 28 August 1918, in *The Battle of St. Mihiel,* 3:57; Lahm, *World War I Diary of Col. Frank Lahm,* 13 September 1918, 126–27.

65. Hugh A. Drum, Supplement No. 1 to Radio Memorandum No. 1, 9 September 1918, in *The Battle of St. Mihiel,* 3:108–9; William Mitchell, Battle Orders No. 1, 11 September 1918, in *The Battle of St. Mihiel,* 3:140.

66. Mitchell, *Memoirs of World War I,* 243.

67. Mitchell, Battle Orders No. 1, in *The Battle of St. Mihiel,* 3:137–40.

68. Mitchell, *Memoirs of World War I,* 241.

69. William Mitchell, Plan of Bombardment Aviation, 17 September 1918, in *The Battle of St. Mihiel,* 3:92–95.

70. Bruce, *A Fraternity of Arms,* 259–63.

71. Hooton, *War over the Trenches,* 247–48.

72. Quoted in Mitchell, *Memoirs of World War I,* 250.

73. Pershing, *My Experiences in the World War,* 272–73.

74. Mitchell, *Memoirs of World War I,* 250.

75. Pershing, *My Experiences in the World War,* 290.

76. Hudson, *Hostile Skies,* 300.

Partners in Name Only

The Royal Air Force and Armée de l'Air during the Battle of France, 1940

Matthew Powell

Relations between Britain and France had been relatively fraught since the end of the First World War and continued to ebb and flow throughout the interwar period.[1] The biggest sources of tension surrounded how both nations should react to Germany's refusal to continue with reparations payments under the Treaty of Versailles of 1923 and how to respond to the actions of Adolf Hitler in his attempts to tear up the treaty and redraw the map of Europe in Germany's favor. At the heart of these tensions was a disagreement in the 1920s over how to impose the treaty and rehabilitate Germany back into the wider European family.[2] French fears over a potentially resurgent Germany added to greater tensions over general European disarmament, with London and Paris disagreeing vehemently over how disarmament should be carried out.[3] While the French felt insecure in Europe, they would never agree to reduce their standing army.[4]

With the rise of Hitler, France, which felt fundamentally weakened due to the increasingly aggressive nature of German foreign policy, looked to a formal alliance with Britain to provide its security against aggression. Such an alliance, however, was difficult for Britain to countenance.[5] Scarred by its experience in the First World War, Britain looked to withdraw from European affairs as much as possible and revert to its pre-1914 position of "splendid isolation."[6] The difficult political relations between Paris and London had a

knock-on effect for relations between the respective nations' military organizations. Talks on a formal alliance, for example, did not take place until the late
1930s and were then hampered by the British policy of Limited Liability and
the French policy of Defense of the West.[7] Despite these difficulties in the early
interwar period, Paris still believed that its security in Europe ultimately rested
on British support.[8] Throughout the numerous political crises of the mid- to
late 1930s, very little discussion and planning took place, and this lack was
highlighted in the performance of both nations' air power in the Battle of
France.[9] These confusions eventually led to a poorly integrated coalition command structure, confusion over the roles of individual air forces, inadequately
trained crews, and, ultimately, disappointing performance on the battlefield.[10]

The Battle of France also demonstrated that the lessons that had been
learned by the Allied powers about the nature and operational tempo of
modern warfare had been misjudged and superseded by the lessons learned
by the Germans.[11] This lack of planning between the declaration of war
and the German invasion of western Europe characterized the "Phony War."
Very little was done to integrate headquarters or command structures or
to improve command-and-control (C2) capabilities. Ultimately, the biggest
restriction on C2 capability was in terms of the numbers of radios available,
particularly for the British.[12]

As the largest contributor to the coalition, the French Army would dictate
Allied strategy. Both partners, however, did share a great deal in common. Both
believed that operations would unfold at a similar tempo to that of the First
World War and that in this respect armored forces were better suited to operating at a similar pace to that of the infantry rather than being used as advanced
"shock" formations to punch holes in the enemy defenses. The French did look
in some ways to modify their operational procedures after German operations
in Poland, but these changes were too little too late and made in almost a blind
panic. They were patchy in nature and for the most part ineffective.[13] Politicians
and senior military leaderships of both nations feared a "knock-out blow" from
the air, and so extensive restrictions were placed on the air efforts. Despite all
of this, however, there was little effort to integrate French and British military
forces prior to the declaration of war in 1939, and nothing could be done on
this issue once war had been declared because the Allies did not want to be
caught out by a surprise attack while their forces were in the process of integrat-
ing their communications systems and force structures, thereby making their
task of defending France and the Low Countries that much harder.

This chapter argues that the seeds of this performance were sewn in the years prior to the Battle of France, and the catastrophic result for the Allies was the reward for poor planning, among a host of other military failures. Ultimately, the defeat of the Allied powers in the battle was in large part due to the overall strategic and operational plan that was put into effect once hostilities began in western Europe. This plan was to move northward and engage German forces at the Dyle River in Belgium.[14] The major factor influencing this decision was the French experience in the First World War, in which a significant proportion of French territory had been occupied, including its northern industrial heartland. The French, therefore, wanted as little of the fighting to occur on French territory as possible.[15] This strategic plan meant that no matter how good or bad the cooperation of the Royal Air Force (RAF) and the Armée de l'air, it was effectively doomed to failure. To support this argument, the chapter explores both the political and the military relations during the interwar period and demonstrates the significant impact they had on the performance of both nations' air forces. It also covers the air operations conducted and how the French C2 system hampered efforts to coordinate air attacks and to react to the overwhelming first blow the Wehrmacht dealt to French forces during its attack at Sedan.

The First World War

The British and French military forces acted in concert to defeat the German Army on the Western Front between 1914 and 1918. Their cooperation, however, was not what would be recognized as a coalition in the post-1945 sense. Both nations' armed forces maintained separate headquarters up to the highest levels of command for the majority of the conflict. A similar position existed between the air forces of each nation. The air wing of the British Army, the Royal Flying Corps (RFC), existed to provide support to British ground forces through contact patrols as well as to contribute tactical air power through close air support and battlefield air interdiction, artillery spotting, and tactical and strategic reconnaissance.

Although there was little in the way of traditional coalition partnerships between the French and British air arms at the higher levels of command, there was a great deal of cooperation in terms of tactical development. The French had a massive influence on the development of British aviation during the First World War, despite the lack of a formal arrangement between the two. One major influence was in the field of aeroengines. A lack of industrial infrastructure meant that there had been little development in this field in Britain, and it was subsequently easier to buy Gnome engines from French manufacturers. Many of the tactical procedures utilized by the RFC in support of the British Expeditionary Force (BEF) were, in addition, based on French ideas. One particularly important aspect revolutionized by French thinking was in "the RFC's approach to aerial photography and map-making."[16] These developments were not based on a formal arrangement but were more ad hoc in nature and relied on the good relations that existed between the two forces at the lower levels of command and operations.

THE INTERWAR YEARS

With the signing of the Treaty of Versailles in 1919, the British, at least believed that large-scale warfare was a thing of the past. France, however, still felt insecure despite the provisions of the peace treaty removing German offensive capability.[17] The huge cost of the First World War in blood and treasure was not, in the opinion of French prime minister Georges Clemenceau as well as of most French politicians, strong enough to prevent the possible resurgence of Germany.[18] The rise of European nationalism and its acceptance by the politicians and diplomats meant that France was strategically weaker than in the pre-1914 period. France's ally, Russia, had succumbed to revolution and was weakened fundamentally by civil war. In the Russian Empire's place, a number of smaller nations had been created. These new states, such as Poland and Czechoslovakia, would not be able to provide the same level of external security to France as Russia had done, and so France turned its gaze across the channel to Britain.[19] France felt that its security was weakened to an even graver degree when Belgium abrogated the Franco-Belgian Alliance due to the passive response of the French over the remilitarization of the Rhineland.[20]

A major flashpoint in relations between Paris and London occurred over the German inability or refusal to pay the reparations due under the Treaty of Versailles.[21] The former Allies did not present a united front with the French and Belgian governments when the latter contemplated sending troops to occupy German industry and extract payment in kind through the seizure of manufactured goods, such as steel and coal.[22] The deterioration in relations that followed this incident led the British government to initiate an air rearmament scheme to deter any potential French action if relations worsened.

Despite the fact that this rearmament scheme was never completed, it serves to highlight the tensions that existed between the two nations, which had been coalition partners only five years earlier.[23] The reality was that the French were never going to be attacking London from the air, but this did not prevent the RAF's Chief of the Air Staff, Air Marshal Sir Hugh Trenchard, from exploiting the general British fear of a knock-out blow from the air. Such an action would have been difficult for French aircraft to undertake. The Armée de l'air was not created as an independent force until 1933, and for most of this period the focus of operations was on the support of ground forces.[24]

As an island nation, Britain was able to develop air power in line with its strategic geopolitical context. With the English Channel to protect it from invasion by land forces, Britain was able to develop a strategic-bombing capability and prosecute economic warfare, which would take several years to be fully effective and impose its will upon a potential enemy. France, similar to Germany, could not simply rely upon strategic bombing as a method of prosecuting a modern conflict.[25] Unlike Germany, which had no real naval presence to concern itself with for the majority of the interwar period, France had to consider its application of air power in two joint contexts: land and maritime as well as an independent approach. This situation was hindered further by the fact that France's air forces were still officially tied to its army and navy, which impeded the development of specialist aircraft.[26]

Ultimately, in an attempt to resolve this issue, the majority of aircraft developed by the French in the early to mid-1930s were multifunctional aircraft capable of conducting a number of roles on the battlefield.[27] Despite the appeal of multirole aircraft, the result was a series of aircraft that were jacks-of-all-trades and masters of none.[28] The decision to develop multifunctional aircraft was to have serious repercussions for the French air force when it came to applying air power during the Battle of France. The lack of air superiority over the decisive parts of the battlefield, when necessary, meant that France's coalition partners would have to take on more of the responsibility with their single-role aircraft.

THE DEVELOPMENT OF AN ALLIANCE

The rise of Hitler and his attempts to change the European order had a significant effect on strategic thinking in France. Effectively isolated on the European continent, France felt its security threatened by the rise of both German and

Italian nationalism. Attempts to act in concert against Germany's demands proved to be fraught with complications. Neither France nor Britain felt it was strong enough militarily to confront Hitler as his demands for territorial appropriation escalated. There were also fundamental disagreements about how best to resolve the political issues that Hitler was creating. Paul Doerr has argued that French proposals were deliberately unrealistic in order to pin the blame for failure on Britain.[29] The French looked to put forward proposals that they knew would be fundamentally unacceptable to Britain and its military priorities.[30] The one thing that tied both Britain and France together, however, was their reluctance to challenge Hitler through a clash of arms as he moved to redraw the political map of Europe.[31] In the aftermath of the Munich Agreement, wherein the Sudetenland area of Czechoslovakia was ceded to Germany without the agreement of the Czechoslovak government, both France and Britain felt that their security was under as great a threat as that which had faced both nations in the summer of 1914.

The British government felt that the long-range and strategic reach of air power, as had been theorized by, among others, Trenchard in 1919, could have a devastating effect on an adversary's population, economy, and infrastructure. In the latter stages of Stanley Baldwin's premiership and under Neville Chamberlain's, the British government saw a large air force capable of conducting attacks on cities and infrastructure as constituting an effective deterrent force.[32] The focus of British rearmament spending was on the RAF and Royal Navy, but the army's role in any potential European strategic defense was not officially settled until early 1939, when it was decided that six divisions of the BEF would support the French Army.[33]

With the eventual commitment of an expeditionary force from Britain to support French operations, staff talks between the two began. These talks were not as fruitful as might have been expected. The enmity of the cool political relations made discussions between the two nations' services difficult. No real effort was made to create a multinational headquarters, and the staff "talks" that did take place were not traditional in nature. They took the form of military missions in which, from the British perspective, a major general or lieutenant general was assigned to French or Belgian headquarters or to the headquarters of army groups or armies. French Army officers who ranked much lower than a major general or lieutenant general and who often were called in from the reserve or from retirement were attached to the headquarters of most British units. For most of these junior to midlevel officers, their role was not to help with the planning of deployments or to discuss how formations would act once hostilities had begun.[34] They were simply there to provide language and cultural bridges to those in the formation to which they were attached.[35]

Although there was a great deal of discussion about the logistics and location of British air forces in France, little was effectively done to integrate them into French C2 structures. Limited information was provided to the British regarding French tactics, either in the air or on the ground, and separate headquarters for British forces were established, just as had been done in 1914.[36] This meant that although allied in a common political cause, the defeat of Nazi Germany, the British and French were two military forces that had nothing to bind them together. They simply acted as distinct and separate entities on the battlefield, with little in the way of joint planning or direction.

Given the European strategic situation in the wake of the First World War, Britain and France did not expect that they would have to combine forces to repel another German attempt at dominating the European continent.[37] To have engaged in such actions in the 1920s and 1930s whereby Britain and France pooled their sovereignty and to a large degree their foreign policy would simply have been unacceptable to their respective publics. Such an arrangement would in theory have been perceived as restricting each nation's ability to act in its own interests and as having the potential to drag it into foreign wars at the behest of an unreliable ally. That both nations arrived at similar conclusions in their thinking about air power can be put down in part to sheer coincidence but also in part to the prevailing strategic conditions and the frame of mind that air power theorists had at that time.[38]

Despite these underlying deficiencies, a War Office pamphlet written for British junior officers provided details of French formations and tactical units but devoted only five pages to French aviation. Originally written prior to closer military ties being established, the pamphlet was updated on a regular basis to provide the most current information possible.[39] Designed for the army, it gave very basic instructions as to how, in theory, French aircraft cooperated with their ground forces, and in many ways it read as if the War Office were jealous that the French had maintained a pre-1914 system that the British were desperate to go back to. There was no detail regarding the tactics of cooperation or how British ground forces were to cooperate with French aircraft if required. The most cogent explanation for this omission is the fact that when the pamphlet was written, the BEF did not have a continental role, and once the Cabinet had actually agreed to this role, there was little time to acquire, process, and disseminate the necessary information. One of the major failings in this respect was that French Army officers were not educated in how to use air power. They had little knowledge of how to use the air assets under their control, and the force structure in which the air assets were deployed meant that they could not be used to great effect. French air squadrons were issued directly to army divisions and placed under their direct operational control—the complete opposite of the system utilized by the RAF. Not only were both nations intending to deploy their forces in structures that were the polar opposite, but it was also almost impossible to integrate them into a wider system that would prove effective for either side or, indeed, both sides. This inability, combined with the French inability to communicate how they would apply air power in support of land operations, meant that coalition discussions and applications of air power would be severely limited.

In addition to the exchange of officers at all levels of command between the two nations, the British also made an overture to the French High Command that a permanent inter-Allied staff should be created to aid the development of plans and provide mutual support prior to the BEF embarking for the continent.[40] The French insisted that this staff be based in Paris, but the British rejected this idea and insisted that it should be based in London. Such an organization would have provided a greater degree of cohesion to the Allied efforts and allowed them to act in concert, but the inter-Allied staff's location in London would have placed a greater strain on France's already unreliable communications network and would have led to delays in decisions being acted upon.

There was agreement between British prime minister Neville Chamberlain and his French equivalent, Édouard Daladier, that a Supreme War Council should be created, similar to that which had existed after the German Spring Offensive of 1918. This council was not to be formed, however, until the outbreak of hostilities.[41] There were still issues with how it would be constituted and function, which would mean it could not be as efficient as its First World War predecessor. Being created only at the outbreak of hostilities meant that that the council's first months would inevitably be confused and that it would be unable to react to developing situations in a timely manner, thus leaving those on the ground in an almost impossible position. The most fundamental failing of the Supreme War Council, however, was in how it was constituted. There was no Allied secretariat to maintain records and preserve continuity in discussion and decision making. The council was also designed to have no executive control over strategic decisions. Final decisions would be taken by governments and not by generals.[42] The efficiency of the decision-making process was not helped by Chamberlain's resignation in May 1940, just hours prior to the German invasion of France.[43]

Despite all of these issues, the delay in operations during the so-called "Phony War" gave the Allies the opportunity to establish a better position for themselves. Unfortunately, they did not do so, and the basic Allied structures required to respond to a German attack were no further advanced come the spring of 1940. A mixture of institutional lethargy, an unwillingness to confront the situation that had presented itself, particularly from a French perspective, and an inability to integrate forces, command structures, and headquarters or to think flexibly about how to prepare defensive doctrine led to a situation where if the initial defensive line fell, any attempt to respond rapidly would be seriously hampered.

PARTNERS AT WAR

The significant time gap between the culmination of German operations in Poland and the beginning of operations in France and the Low Countries should have seen a substantial rise in coalition training, exercises, and integration.[44] But the French High Command placed many restrictions on the training that the RAF could conduct. RAF training operations had to refrain from crossing the Franco-German border lest the Germans use any accidental breach as a pretext for invasion.[45] Little work was conducted to integrate the two air headquarters, nor was there any thought to colocating air and ground headquarters at any level of command on either a national or a coalition basis,[46] despite the importance of colocated headquarters in aiding the ability of aircraft to support ground troops at both the tactical and operational levels of war.[47] The RAF, split into two separate formations, the Advanced Air Striking Force (AASF) and the RAF Component of the BEF, sought French permission to conduct exercises to improve its capabilities. The French High Command agreed to this request but stated that the training could not take place near the border. Perpignan was suggested as a suitable location where the exercises could be accommodated.[48] Perpignan, however, did not have the same terrain as Belgium or northern France. Although tactics and procedures could be tested, there was little opportunity for pilots to become accustomed to the topography in the area where they would be providing support for land forces in active operations against a formidable opponent.

The RAF had also suffered innumerable problems in the development of tactical air power throughout the interwar period, although the RAF had advanced to a far greater state in a theoretical sense than the Armée de l'air or than has been acknowledged in the literature to date.[49] That is not to say that there was agreement between the Air Ministry and the War Office over how to employ air power in support of ground forces. This was yet another complicating factor in the utilization of air power during the Battle of France.

The command system under which the two nations would operate in the event of a German invasion was clear inasmuch as the hierarchy had been defined. The BEF, under General Lord Gort, was under the nominal command of the French commander in chief, General Maurice Gamelin. Although there was an agreed overall strategy, neither the RAF nor the Armée de l'air was designated to work with the other. In fact, after the large reorganization of the RAF in France, where the AASF and BEF Component were joined together to form the British Air Forces in France (BAFF), it was made clear to their commander, Air Marshal Sir Arthur "Ugly" Barratt, that he was not under the operational control of any French commander, as Gort was by this time.[50]

Over the autumn and winter of 1939–1940, the lack of operations and general inactivity of the French forces in particular led to a significant drop in morale. That winter was one of the coldest in living memory, and troops were set mundane tasks such as defensive fortification building and watches.[51] The French received very little in the way of training because their military leadership had staked their success on the newly constructed Maginot Line to protect them from any German onslaught, and morale was generally low due to political unrest and poor pay.[52] A series of defensive trench systems supported by the fortifications of the Maginot Line, similar to those constructed during the First World War, were dug, from which it was expected that the infantry would hold off German attacks.[53] This change in operational and strategic-level outlook forced a radical change in strategic thinking from the French High Command. The infantry would still utilize the Maginot Line to protect the majority of the French frontier, but they would advance into Belgium once hostilities had begun and prepare defensive positions alongside natural defenses.

Two feasible options were open to the French with this strategy. One was to position themselves along the Escaut (Schelde) River or to advance to the Dyle River. Both options would provide a natural defensive position that could be fortified in places if necessary and if time allowed. These fortifications would, however, be difficult and costly to build due to the high-water table of this particular region of Belgium. Any fortifications would have to be built upward rather than dug into the ground, adding to cost and the time necessary to build them.[54] The fortifications were not completed by the time the Second World War broke out and so could not be factored into a wider Allied strategy. This difficulty was further heightened by the facts that French training standards were generally poor and that little time was dedicated to antiaircraft training or antitank defenses.

ALLIED AND GERMAN WAR PLANS

The French (later Allied) strategic plan was based around expectations of where Germany would attack from and how operations would unfold. The expectation was that the German Army would seek to exploit the vast space of Belgium and the Netherlands in a fashion similar to the Schlieffen Plan of 1914. French intelligence estimates of German planning were initially correct because this was exactly what the Germans had planned originally.[55] Appearing to learn the lessons from operations in 1914, however, the Germans decided they would instead swing down through the Netherlands and Belgium, utilizing the extra space the Netherlands would give the German Army to accommodate a greater number of troops. The German Army would also feel more secure on its eastern flank than it did in 1914 due to its successful invasion of Poland and the Molotov-Ribbentrop Pact with the Soviet Union.

The Dyle plan sought to hold the Germans along the river in operations similar in style and pace to those of the First World War from 1915 to 1917. Slow operations, dominated by the defense and capitalizing on the operational-level pace and agility that railways provided, allowed gaps to be plugged before the infantry could exploit any breakthrough in depth. The infantry was able to cross No Man's Land only on foot and had to cover broken ground while facing heavy defensive fire.[56] Despite improvements in the power and reliability of the internal combustion engine since 1918, the French still viewed the tank as an infantry-support weapon, and so they assigned armored formations to infantry in penny packets, thus reducing the formations' ability to provide overwhelming firepower and pace and agility at the tactical level.[57] After a degree of experimentation in armored formations, the British reached a similar conclusion on how to deploy tanks in battle.[58]

The original German invasion plans were inadvertently turned over to Belgian authorities after the officer carrying them crash-landed in Belgium on 10 January 1940 after his pilot got lost. This loss of vital intelligence prompted a radical change in German operational planning, which ultimately gravely undermined French defensive plans.[59] Allied air forces were to support the advance into Belgium, and the RAF was to look to attack strategic targets in the German homeland because the Armée de l'air did not have the capacity.[60]

OPERATIONS

There had been a number of false alarms concerning a German invasion from the end of operations in Poland throughout the winter and early spring of 1940, and the fact that they were false led to a feeling of complacency.[61] This was particularly the case within the French High Command. The French High Command also suffered from a series of intractable problems, most of which were of their own making. Due to their doctrinal thinking on the pace of operations, they continued with a top-down, hierarchical, command system, which required requests to be passed up the chain of command for approval before being passed back down using the French General Post Office telegraph system rather than radio.[62] This command system gave very little flexibility to junior officers on the ground to react to situations, and the little training they received after being conscripted inhibited them even further. The C2 system employed to facilitate this command arrangement was slow and cumbersome, and the head of the French Army, General Gamelin, did not even have a telephone in his headquarters, preferring to rely on dispatch riders using motorcycles to carry messages to those below him in rank. It was into this structure that the RAF had to fit, although it continued to utilize its own C2 system. This system was, however, susceptible to overload at key times and liable to fail in an emergency.[63]

With advanced news of a possible German invasion, the Allied forces moved quickly to initiate Plan D and accomplished their advance to the Dyle with very little hinderance.[64] The RAF was able to conduct advanced reconnaissance of the potential area of operations and supported the BEF effectively as it engaged with the German forces advancing through Belgium. This advance was, however, a diversionary attack designed to take the Allied focus away from the key sector of German operations, the Ardennes Forest, Sedan, and the Meuse River. The Luftwaffe followed a simple two-step plan in its support of the German Army. First, it would saturate the area with air power and look to seal off the battlefield with air interdiction operations. These operations consisted of attacking aircraft on the ground and disrupting command, control, and communication systems to gain localized air superiority. Although both the Armée de l'air and the RAF lost relatively few aircraft through Luftwaffe interdiction operations, attacks were concentrated around where the German Army sought the *Schwerpunkt* (decisive point) of the operation, and through this concentration the Luftwaffe gained almost complete control of the air around the Meuse River at Sedan, Namur, and Dinant.[65] Once it had achieved this, it would then support ground forces with a mixture of close air support missions, reconnaissance, and advanced flank protection.

A myth has emerged, however, about how the Luftwaffe deployed air power in the Battle of France: that the Germans were overly reliant on close air support for their success in the initial phases of their assault across the Meuse.[66] Yet only one crossing of the Meuse required close air support; the others were achieved with no air support at all.[67] Although the Luftwaffe could conduct close air support missions with a degree of skill and flair, it was far more effective at gaining localized control of the air and using this domination of the skies to conduct interdiction missions.

As the actual focus of German operations dawned on the Allies, their attention turned quickly. They were concerned with reports of a large German breakthrough at Sedan and the flow of armored and mechanized traffic across the river. Both air forces conducted large-scale reconnaissance of the region. Initial reconnaissance conducted by the RAF around the Ardennes revealed large tailbacks of German forces that stretched all the way back into Germany itself.[68] This reconnaissance was conducted by the RAF because even at this initial stage of hostilities the Armée de l'air was struggling, due to the speed of the German advance, to maintain its forward airfields so that it could be an effective support for the French Army. The French High Command quickly dismissed what the RAF's reconnaissance revealed as nothing more than a diversionary attack because they were still clinging to their belief that the Ardennes was impassable for large armored formations.

Barratt asked permission from the French High Command to attack the advancing German columns in an attempt to cut the head of the snake from the body and seal off the battlefield. This would mean that any initial break-through made by the Germans could not be supported, would run out of momentum, and be pushed back easily. Permission to attack the advanced columns was denied to Barratt and his forces, however, because it was felt that if the bombers sent on the mission were to miss the military targets and hit civilians by mistake, it could result in large-scale retribution attacks on French towns and cities. This was a risk that the French High Command were unwilling to take.[69] This fear of German retribution prevented BAFF from conducting a mission that could have altered the outcome of the Battle of France and with it the conduct of the Second World War. This decision was a prime example of coalition air power being ineffective and lacking flexible strategic direction to react to an unexpected and unfolding situation. Despite being refused permission, Barratt realized the strategic importance of these convoys and eventually set about attacking them despite the refusal.[70] He conducted these attacks too late in the course of the battle and with too little force, however, to have any meaningful effect on the German advance.

The Allies were slow at realizing that the decisive point of German operations was in and around Sedan and the Meuse River. Once they realized this, they were slow to stem the tide of the German advance. They had fallen foul to a German military dictum that stated, "An error in the first stages of deployment can never be made good."[71] The Allies were unable to redeploy their armies in enough strength or time to be able to mount significant counterattacks and had to rely on air power to accomplish this and stop the German advance in its tracks. The decision to conduct this operation was taken a full two days after the initial German crossing of the Meuse, which began on May 12 and allowed the Germans to break out and advance through open country. The advanced Panzer formations were also reliant on the support being transferred along the bridgeheads that had been established by German engineers. If these bridgeheads could be cut off or destroyed, both the tactical and the operational mobility of the Panzers would be severely restricted.[72] A coalition meeting took place between the senior British and French land and air commanders on 13 May. It was agreed that the Armée de l'air's and BAFF's medium bombers would attack the pontoon bridges on 14 May. The BAFF would attack in the morning, and the Armée de l'air would follow up the attack in the afternoon.[73]

The RAF had already investigated the tactics necessary in attacking targets such as bridges and had reached the conclusion that it was extremely difficult, if not impossible, to achieve direct hits from medium altitude and higher. Flying at this altitude, however, was deemed necessary to get aircraft to the targets.[74] The situation by 13–14 May was so serious that there was little option for both air forces. Even by this early point in the battle, the defense of France was hanging in the balance. If the Germans were allowed to continue their advance unabated, then the forces that had moved to the Dyle would quickly become redundant, and the Germans would be presented with a tantalizing option: to advance to the French coast or turn and attack Paris.

With the attacks scheduled, both French and British forces made preparations to ensure they would be as successful as possible. The Allies were not the only ones to realize the importance of the pontoon bridges spanning the Meuse. This was the only arterial route available to the Germans, and any future success would rely on its protection. German fighter aircraft had been stationed nearby, and a vast number of antiaircraft guns had been brought up to harass and prevent any attack. The morning attack conducted by the BAFF had very little impact, with only one bomber being able to confirm a successful hit on a bridge. The bombers (Fairey Battles) attacked in four waves and at three-hour intervals, accompanied by fighter escort. The same tactics were to be used for the attacks in the afternoon, but these later attacks again fell to the BAFF because the Armée de l'air was unable to find enough serviceable aircraft. Knowing that the German antiaircraft defenses would be on high alert after the morning attacks and that the afternoon forée was in effect a suicide mission, Barratt informed the Battle crews scheduled for duty that the operation was on a volunteer-only basis.[75] Each member of the squadron stepped forward for duty, but only thirty-one of the seventy-one aircraft that conducted the attack returned.[76] The lack of coordination during this attempted counterattack of the German thrust was due to several factors, one of which also leads to an explanation of why the French felt unable to conduct the afternoon attacks and why the responsibility for them again fell to the BAFF.

The French felt that the location of the German attack, the Ardennes Forest, was impassable to heavy and cumbersome armored forces with mechanized support formations, and so they paid very little attention to any German movements in that area. There was no overall plan for conducting any offensive or defensive operations in the region, and so it was held only by third-rate conscripts, who were highly unmotivated and of a general poor quality as soldiers. Despite the region's historic significance resulting from previous decisive battles between various German states and France in previous centuries, little aerial or ground reconnaissance of it was conducted, so detailed information to identify targets that could be attacked or ways of preventing an enemy attack being reinforced and supplied through battlefield air interdiction operations was not available. The Armée de l'air was also in no position to have any real positive effect on the battlefield, a problem that can be traced back to the difficult birth of independent air power in France as well as to the sheer doctrinal confusion that led its air force to this situation. There was no real or clear demarcation between acting in support of French land forces or acting in a truly strategic capacity. This lack of clear doctrinal direction for the Armée de l'air in the interwar period, combined with the French Army's understanding of air power, meant that there was a great deal of confusion in the development of aircraft technology.

This understanding of air power was a major factor behind the development of the battleplane concept: an aircraft that could conduct a number of different roles across the battlefield and beyond into enemy territory in a strategic capacity. Unfortunately, such an aircraft proved to be too heavy and cumbersome to achieve success in even the most basic of battlefield tasks and so was utilized mainly as a strategic bomber. Even in this role, however, it struggled to have a decisive impact because its overall weight without ordnance meant that it could never carry a sufficient weight of bombs. The general speed of French aircraft, combined with the Luftwaffe's attacks prior to the offensive at Sedan, meant that they were easy prey and could provide no real decisive effort when required to intervene. The Wehrmacht had also been given the time to create a deadly defensive force of fighter aircraft and anti-aircraft fire.[77] This was the situation that confronted the Armée de l'air when the counterattack at Sedan was planned. It is also arguable that the French knew in advance that they did not have either the numbers or right types of aircraft required to conduct such an operation and had agreed to conduct the afternoon attack at Sedan only as way of keeping the RAF and, by default, the British government onside.

Conclusion

The attacks at Sedan effectively ended the BAFF's role in Allied air power in any effective sense. With the Armée de l'air struggling to keep aircraft operational, the attacks also spelled the end of Allied air power being able to intervene in German actions as they continued their advance and conquered France in six weeks. This quick domination came as a huge shock not only to the French themselves but also to the wider world, who for most of the interwar period saw the French Army as the most powerful in the world. The Luftwaffe continued to exploit the lack of Allied air power, using its aircraft to conduct reconnaissance in depth and acting as a flank guard as advanced German formations made rapid progress toward the English Channel coast.

As the situation in France deteriorated and the Armée de l'air found it almost impossible to respond at the decisive point of operations around Sedan, on 14 May, it requested that a further six squadrons of British fighters be sent to France to aid in the planned attempts to stem the flow of German armored and mechanized formations and to potentially launch a counterattack. The British response, led by the air officer commanding in chief of Fighter Command, Air Marshal Sir Hugh Dowding, was an outright refusal, and this decision has caused tension between Britain and France ever since.[78] In refusing to send extra fighter squadrons to France, Dowding had two concerns. After seeing the disaster unfolding in front of his eyes and the loss rates being inflicted on the various BAFF aircraft already stationed in France, he considered that the situation was completely lost and that he would simply be sending squadrons to be needlessly slaughtered while having no real effect against the Germans. The other factor that must have been playing on Dowding's mind as he announced his decision to the War Cabinet, which had been summoned in part to consider this request, was that with the writing on the wall for France, it was inevitable that Britain would be next in the Germans' sights. This next battle would be one of fighter against fighter, and so a careful husbanding of resources was required to prevail in the future. Britain, Dowding in particular, could not afford to fritter precious fighters away on the lost cause that the Battle of France appeared to be.

Among the recriminations of the Battle of France, the British Army used the issue of tactical air power as a stick with which to beat the RAF and attempt to make it take its development more seriously. The poor performance of both the French and British, acting alone and in concert, can be put down to a number of factors: superior military thinking and skill on the part of the Germans, poor operational and strategic planning on the part of the Allies, and simple bad luck that the plans of both adversaries meshed so well for one of the first times in history. The lack of British and French interaction during the Battle of France was, however, very much the result of the twenty years of difficult relations between them, caused largely by a misunderstanding of how both perceived the changing European situation of the 1930s and how their individual security had become intertwined.

The lack of a common policy in the face of the rise of Hitler, with both nations seeking to blame the other for any possible failure, meant that Allied staff talks were not as fruitful as they could have been. With talks not being conducted before the Munich Agreement, there was little time available for the army in Britain, unsure of its role prior to early 1939, to develop a clear line of strategic thinking. The outcome of the Battle of France in 1940 provided both nations with a scapegoat it could blame for the ultimate failure to defeat the Wehrmacht in this early stage of the war, which would sour relations between them in the years after 1945. The French could point to the RAF and their refusal to send more fighters in support of the beleaguered French counterattacks, despite ending the campaign with more aircraft than were available in May. The British could highlight the poor strategic plan that was put into action, which constrained what could be done once the Germans had launched the plan devised by Generalleutnant Erich von Manstein and refined by Lieutenant General Heinz Guderian. Allied air power did fail in 1940, but the biggest failure was that of long-term coalition planning and strategic thinking.

Notes

1. The author thanks Dr. Jamie Slaughter for his helpful and insightful comments on a previous draft of this chapter.

2. Gary Sheffield, *Forgotten Victory: The First World War, Myths and Realities* (London: Review, 2002), 272.

3. Maurice Cowling, *The Impact of Hitler: British Politics and British Policy* (New York: Cambridge University Press, 1975), 65.

4. Carolyn J. Kitching, *Britain and the Problem of International Disarmament, 1919–1934* (London: Routledge, 1999), 15, 59.

5. Martin S. Alexander and William J. Philpott, introduction to *Anglo-French Defence Relations between the Wars,* ed. Martin S. Alexander and William J. Philpott (Basingstoke, UK: Palgrave Macmillan, 2002), 2.

6. Andrew David Stedman, *Alternatives to Appeasement: Neville Chamberlain and Hitler's Germany* (London: I. B. Tauris, 2015), 16.

7. Talbot Imlay, "France, Britain, and the Making of the Anglo-French Alliance, 1938–1939," in *Anglo-French Defence Relations between the Wars,* ed. Alexander and Philpott, 101.

8. Dick Richardson, *The Evolution of British Disarmament Policy in the 1920s* (London: Pinter, 1989), 12–13.

9. Hew Strachan, *European Armies and the Conduct of War* (London: Routledge, 1983), 166.

10. "The Campaign in France and the Low Countries: September 1939–June 1940—Part I, before the War: The Anglo-French Conversations," 1, AIR 41/21, The National Archives of the UK (TNA), London.

11. Karl-Heinz Frieser, *The Blitzkrieg Legend: The 1940 Campaign in the West* (Annapolis, MD: Naval Institute Press, 2005), 348.

12. Edward Smalley, "Signal Failure: Communications in the British Expeditionary Force, September 1939–June 1940," *War in History* (forthcoming), currently available at https://journals.sagepub.com/doi/full/10.1177/0968344518804846.

13. Brian Bond, "Introduction: Preparing the Field Force, February 1939–May 1940," in *The Battle for France and Flanders: Sixty Years On,* ed. Brian Bond and Michael D. Taylor (Barnsley, UK: Leo Cooper, 2001), 7.

14. "The Campaign in France and the Low Countries," 155–56.

15. Joe Maiolo, *Cry Havoc: The Arms Race and the Second World War, 1931–1941* (London: John Murray, 2010), 82.

16. Peter Dye, *The Bridge to Airpower: Logistics Support for Royal Flying Corps Operations on the Western Front, 1914–1918* (Annapolis, MD: Naval Institute Press, 2015), 26–27.

17. Richardson, *The Evolution of British Disarmament Policy in the 1920s,* 7.

18. Ibid., 6–7.

19. P. M. H. Bell, *The Origins of the Second World War in Europe,* 2nd ed. (Harlow, UK: Longman, 1997), 26.

20. "Notes on Strategy and Tactics, Part I: Strategy," chapter 16 in *Handbook on the French Air Forces,* May 1939, AIR 10/648, TNA; Ronald E. Powaski, *Lightning War: Blitzkrieg in the West, 1940* (Edison, NJ: Castle Books, 2003), 8–9.

21. Stephen Budiansky, *Air Power from Kitty Hawk to Gulf War II: A History of the People, Ideas, and Machines That Transformed War in the Century of Flight* (London: Viking, 2003), 4–5.

22. James P. Levy, *Appeasement and Rearmament: Britain 1936–1939* (Lanham, MD: Rowman and Littlefield, 2006), 4.

23. John Terraine, *The Right of the Line: The Royal Air Force in the European War 1939–1945* (London: Sceptre, 1985), 10–11.

24. Anthony Christopher Cain, *The Forgotten Air Force: French Air Doctrine in the 1930s* (Washington, DC: Smithsonian Institution Press, 2002), 11–14.

25. Williamson Murray, "British and German Air Doctrine between the Wars," *Air University Review* 30, no. 3 (March–April 1980), at https://www.airuniversity.af.edu/Portals/10/ASPJ/journals/1980_V0131_N01-6/1980_V0131_N03.pdf.

26. Cain, *The Forgotten Air Force,* 2.

27. Pascal Venneson, "Institution and Air Power: The Making of the French Air Force," in *Airpower: Theory and Practice,* ed. John Gooch (London: Frank Cass, 1995), 49.

28. Robert J. Young, "The Strategic Dream: French Air Doctrine in the Inter-war Period, 1919–1939," *Journal of Contemporary History* 9, no. 4 (October 1974): 60.

29. Paul W. Doerr, *British Foreign Policy 1919–1939* (Manchester, UK: Manchester University Press, 1998), 221–22.

30. Kitching, *Britain and the Problem of International Disarmament,* 58–59.

31. A. J. P. Taylor, *The Origins of the Second World War* (London: Penguin, 1964), 102.

32. Sebastian Ritchie, *Industry and Air Power: The Expansion of British Aircraft Production, 1935–1941* (London: Routledge, 1997), 41.

33. Alistair Byford, "Fair Stood the Wind for France: The Royal Air Force's Experience in 1940 as a Case Study of the Relationship between Policy, Strategy, and Doctrine," *Air Power Review* 14, no. 3 (Autumn–Winter 2011): 43.

34. Reply to the note Verbale submitted by the British air attaché to the chief of the 2nd Bureau of the French Air Staff, 10 September 1938, AIR 40/2032, TNA.

35. Peter Caddick-Adams, "Anglo-French Co-operation during the Battle of France," in *The Battle for France and Flanders,* ed. Bond and Taylor, 36.

36. Provisional instructions to Air Commodore F. P. Doe, 26 August 1939, AIR 35/3, TNA.

37. Doerr, *British Foreign Policy,* 93.

38. Compare Young, "The Strategic Dream."

39. Notes on the French Army, November 1937, WO 287/17, TNA.

40. "The Campaign in France and the Low Countries: September 1939–June 1940," AIR 41/21, TNA.

41. William J. Philpott, "The Benefit of Experience? The Supreme War Council and the Higher Management of Coalition War, 1939–1940," in *Anglo-French Defence Relations between the Wars,* ed. Alexander and Philpott, 209.

42. Caddick-Adams, "Anglo-French Co-operation," 36–37.

43. Brian Bond, *The Pursuit of Victory: From Napoleon to Saddam Hussein* (Oxford: Oxford University Press, 1996), 152.

44. John Buckley, "The Air War in France," in *The Battle for France and Flanders,* ed. Bond and Taylor, 117.

45. Kate Caffrey, *Combat Report: The RAF and the Fall of France* (Swindon, UK: Crowood Press, 1990), 17.

46. Matthew Powell, "Debate, Discussion, and Disagreement: A Reassessment of the Development of British Tactical Air Power, 1919–1940," *War in History* (forthcoming), currently available at https://journals.sagepub.com/doi/abs/10.1177/0968344519829787.

47. Matthew Powell, *The Development of British Tactical Air Power, 1940–1943: A History of Army Co-operation Command* (Basingstoke, UK: Palgrave Macmillan, 2016), 16.

48. Headquarters BAFF to Headquarters AASF—Night Flying Training, 22 January 1940, AIR 24/681, TNA.

49. Powell, "Debate, Discussion, and Disagreement."

50. Minister for the Co-ordination of Defense Part I, memorandum, 2 December 1939, CAB 66/3/27, TNA; "The Campaign in France and the Low Countries: The Formation of British Air Forces in France," AIR 41/21, TNA; Order of Battle of the AASF, n.d. (c. 1939), AIR 24/679, TNA.

51. Bond, "Introduction," 7.

52. Philip Warner, *The Battle of France, 1940* (London: Cassell Military Paperbacks, 1990), 10; Powaski, *Lightning War,* 22.

53. Alistair Horne, *To Lose a Battle: France 1940* (London: Papermac, 1990), 84–85.

54. Ernest R. May, *Strange Victory: Hitler's Conquest of France* (London: I. B. Tauris, 2000), 288.

55. Ibid., 227–29.

56. Georges Blond, *The Marne: The Battle That Saved Paris and Changed the Course of the First World War,* trans. H. Eaton Hart (London: Prion Books, 2002), 97–132.

57. Robert A. Doughty, "The French Armed Forces, 1918–1940," in *The Interwar Period,* vol. 2 of *Military Effectiveness,* new ed., ed. Allan R. Millett and Williamson Murray (New York: Cambridge University Press, 2010), 57.

58. Brian Bond, *British Military Policy between the Two World Wars* (Oxford: Clarendon Press, 1980), 129.

59. May, *Strange Victory,* 316.

60. "Employment of British Heavy Bomber Force in Support of Land Operations in the Event of a German Invasion of Belgium," 20 January 1940, AIR 35/155, TNA; draft translation from Commander in Chief French Air Force to British Air Forces in France, "Instruction for the Organisation of the Command of Close Support, Bomber, and Fighter Air Forces in Combined Operations," 21 March 1940, AIR 35/3, TNA.

61. Richard K. Betts, "Surprise despite Warning: Why Sudden Attacks Succeed," *Political Science Quarterly* 95, no. 4 (Winter 1980–1981): 553.

62. Peter Caddick-Adams, "The German Breakthrough at Sedan, 12–15 May 1940," in *The Battle for France and Flanders,* ed. Bond and Taylor, 14–15.

63. David Ian Hall, *Strategy for Victory: The Development of British Tactical Power, 1919–1943* (Westport, CT: Praeger, 2008), 51.

64. Jeffrey A. Gunsberg, "The Battle of the Belgian Plain, 12–14 May 1940: The First Great Tank Battle," *Journal of Military History* 56, no. 2 (April 1992): 208, 217.

65. Buckley, "The Air War in France," 123.

66. Ibid., 111.

67. Williamson Murray, "The Luftwaffe Experience," in *Case Studies in the Development of Close Air Support,* ed. Benjamin Franklin Cooling (Washington, DC: Office of Air Force History, 1990), 92.

68. Powell, *The Development of British Tactical Air Power,* 62.

69. "BAFF Barratt's Dispatch, Part II: Work of BAFF Prior to the Land Battle," July 1940, AIR 35/354, TNA.

70. "The Campaign in France and the Low Countries September 1939–May 1940, Military Summary," 10 May 1940, AIR 41/21, TNA.

71. Powaski, *Lightning War,* 44.

72. Bryan Perret, *A History of Blitzkrieg* (New York: Stein and Day, 1982), 94.

73. L. F. Ellis, *The War in France and Flanders* (Uckfield, UK: Naval and Military Press, 2004), 55.

74. "Notes on the Attack of Armored Fighting Vehicles and Mechanical Transport on Roads and the Attack of Railways," 1 November 1939, AIR 14/107, TNA.

75. Robert Jackson, *Air War over France 1939–1940* (London: Ian Allen, 1974), 54.

76. Victor F. Bingham, *Blitzed: The Battle of France May–June 1940* (New Malden, UK: Air Research Publications, 1990) 64.

77. Horne, *To Lose a Battle,* 389.

78. "Conclusions of Cabinet," Minute 2, 15 May 1940, CAB 65/13, WM (40), 123rd Meeting, TNA.

Creating an Anglo-American Air Component

The Western Desert Experience, 1940–1942

Andrew Conway

By the end of 1942, the British-led Western Desert Air Force (WDAF) incorporated a significant US Army Air Forces (USAAF) presence and relied extensively on American-built aircraft to conduct operations. The basis of this metamorphosis was multifaceted and reflected Anglo-American cooperation across political and logistical as well as operational spheres. Any analysis of the multinational dynamics of Anglo-American air power in the Western Desert must therefore encompass this wider context. A key consideration for much of this period was the US transition from nonbelligerent to silent partner to active combatant. Accordingly, this chapter highlights issues of diplomacy, organization and supply, and command and control, alongside the strategic, operational, and tactical implications of these issues for the nascent Anglo-American partnership. It also addresses how these interrelated areas ultimately combined to promote and then crystallize Allied air operations in the Western Desert by mid-1942. Finally, it demonstrates that resolving the tensions and embracing the opportunities arising from multinational engagement often hinged on the interaction occurring at various levels and between various agencies, both military and civilian.

Early Momentum for Multinational Air Operations

The possibility of employing USAAF units in the Middle East was evident from the first American observer missions to the region in 1940. Indeed, it informed part of the larger rationale behind their dispatch and acquisition of in-depth knowledge of Royal Air Force (RAF) procedures and practice. As Major General George Brett, chief of the US Army Air Corps, acknowledged to General Henry Arnold, head of the USAAF, shortly after his arrival in the region in September 1941: "In all my investigations and questions I keep constantly in mind, without mentioning it, the fact that if the United States enters this war, this may be the theater in which they will operate. With this in mind I try to visualize exactly what we would do, how we would do it, where we should be located and all special jigs, fixtures and paraphernalia which we might need and which the British have learned through bitter experience."[1]

Brett's perspective and emphasis formed part of a wider continuum. A succession of American observers and missions stationed in the theater aired their thoughts regarding the possibility of USAAF operations there. Through late 1940 and early 1941, in evaluating RAF Middle East's performance in the Western Desert, Air Corps colonel Gerald Brower considered the prospects quite favorable. In November 1940, he advised Arnold that long-range bombers were "desperately" required in order to reach the principal Italian-held ports at Benghazi and Tripoli.[2] The following month he wrote that American-built heavy bombers would be an invaluable addition to the British air arsenal, commenting that "range for bombers is all-important—think what they could do here with B-17s and B-24s!"[3]

Such were the logistic demands placed on operations in the Middle East generally that Brower concluded any future joint venture between the respective air forces would have to accommodate this reality. As he informed Arnold in January 1941, "I fear it would be impracticable and undesirable in many ways to have our air units; staffs-squadrons-and bases; mixed up in the same area with the RAF. . . . Base areas of the two forces should be well separated." Nevertheless, there were promising signs that the two air forces could work together in operational unison. Here Brower foresaw how "squadrons of both forces could readily use fields and bases of the other, for advanced refueling points; and could use common advanced grounds for short periods." Ultimately, although operating from well-separated base areas, "both forces could concentrate readily, whenever desired, on targets and objectives within range of either. At the same time, they would retain the advantages of dispersion at their bases."[4]

The suggestion of employing American heavy bombers and even American heavy-bomber units became more popular as 1941 progressed. Content to focus his attention on examining logistical issues in the theater, the chief purpose of his visit, Brett periodically commented on possible future operations. Expressing his belief that long-range bombers should be deployed to the theater and launched against Axis ports and supply lines, he reported to Arnold that B-24s would be a suitable choice in light of the targets and distances involved.[5] Alongside his efforts in improving maintenance facilities and the channels of communication between theater and national supply authorities, Brett also took several concrete steps to encourage the deployment of both American-built heavy bombers and USAAF personnel to the Middle East. In late September, he urged Arnold to consider diverting to the theater the American aircraft allocated to the United Kingdom and Russia. In particular, he recommended that twenty-four B-24s currently destined for the British Isles be sent to RAF Middle East instead, where they could be most profitably employed on day and night operations approximately one week after arrival. His case, Brett argued, was not based on sympathy but on a firm belief that "it will bring about consolidation in Middle East and the local political situation and offer, instead of a continual defensive plan, the one possibility of definite offensive action."[6] Although Arnold was initially resistant to the idea, Brett persisted, subsequently winning over British authorities in London; on 24 October, he advised Washington that sixteen Liberators would be turned over for delivery to Air Marshal Arthur Tedder, air officer commanding in chief (AOC-in-C), RAF Middle East Command, with additional spares and supplies to be sourced from the United States.[7] As if to underline this commitment, Brett also thought it essential that an air attaché be assigned to the region as well as officers from the Air Force Combat Command to "observe and report on tactical air operations in the Middle East." "Operations in this theater differ from any yet observed," he informed Arnold, "and we should be getting the benefit of the experience."[8]

In late November 1941, US president Franklin Roosevelt dispatched one of his personal representatives, William Bullitt, former ambassador to France, to the theater. Ostensibly there to report on and assist the British in matters of supply, in a series of telegrams Bullitt gave strong backing to the efforts of theater ground and air forces. Rather than countenancing any diversion of forces and material from the region (a policy Bullitt considered "folly"), he thought they should be strengthened to the best of America's ability. Above all, it was "vital" and "essential" that scheduled deliveries of American-built aircraft, their equipment and ammunition, and the projects initiated by Brett should continue.[9] Following talks with Tedder, Bullitt passed on Tedder's plan to secure three squadrons of Liberators to attack the Ploesti oilfields in Romania. The AOC-in-C Middle East, he informed Roosevelt, was "most insistent that I should present this possibility to you as the vital link in the whole strategic picture."[10] On his return to Washington, Bullitt "talked 'Middle East' up and down" with Field Marshal Sir John Dill, chief of the British Joint Staff Mission, although the latter doubted that Bullitt's enthusiasm would rub off on the American Joint Chiefs of Staff, who "never fully appreciate the activities of unrelated free-lances!"[11]

A powerful influence was also exercised on Bullitt's thinking by American military personnel stationed in the theater. Brigadier General Russell Maxwell, head of the US Military North African Mission, and Lieutenant Colonel Elmer Adler, chief of its Air Section, repeatedly emphasized to the presidential envoy the theater's strategic importance. In December 1941, they also arranged a conference with Bullitt, Tedder, and senior RAF supply representatives to clarify the future prospects of American combat involvement in the Middle East. Adler recounted how in the "very interesting" discussions that followed he received the impression that the British "considered our organization more or less as a supply aiding committee which was justifiable under the Lend Lease concept." But given that America had now become an active belligerent, he informed the British representatives that "it was essential that we be considered in the command sense." With the subsequent placement of Maxwell and Adler on the British Middle East War Planning Council, Adler felt that another significant step forward had been made: "We have broken through the crust of the British reserve and are being admitted now on an equal basis instead of in some subordinate capacity."[12]

Adler continued to be a firm proponent of launching a major Anglo-American military campaign in the Middle East. In a series of strategic appreciations prepared for Arnold and the US War Department, he compared the theater favorably with the alternate prospect of offensive action in the Far East. Despite the logistic complications involved, Adler could see "no reason why the ground supply lines, supported by air transport in larger quantities than we have ever dreamed of previously, could not support an air force in this area." "With such power in place," he argued, "it would be possible to start a systematic clean-up of the Mediterranean."[13] Adler believed that "an air force of any size can be supported in northern Africa, Iran, and Iraq," which would bring "the vital target of Germany within reach of bombardment."[14] Hence, the deployment of American arms would reap the logistic benefits of opening the Mediterranean shipping lanes and facilitate linking up with British and Russian forces to assault Germany from the south.[15]

By February 1942, Adler thought that what had been achieved in the theater thus far clearly facilitated the future employment of American air units. With various US-sponsored aircraft, maintenance, and repair projects under way, he informed Arnold, all that remained would be to provide combat units with their own advanced mobile servicing crews, "and Service Command organization would be complete." Indeed, extension of USAAF combat operations to the Western Desert seemed a necessary and preferable prelude to undertaking an amphibious landing in Algeria or Morocco, a prospect recently discussed during Anglo-American staff talks in Washington.[16] Adler also advocated that a "Bomber Command" be organized in the region during this period. "I have done considerable work in preparing the way for that organization," he informed Brett, "and find that it meets with enthusiastic support from everyone." Adler thought that such an organization, reflecting the cumulative efforts by not only himself and Brett but also by the strong vanguard of American technical personnel who had already visited the theater, could "be completed rapidly in the area assigned and moved to such point as you might desire to fit your battle plans."[17]

THE EVOLVING STRATEGIC CONTEXT AND EMERGING OPERATIONAL IMPERATIVES

The prospect of employing American fighting forces in the Middle East, particularly those of the USAAF, strengthened as 1942 progressed. Although the weight and urgency of commitments to the Pacific theater and long-term strategic planning for a cross-channel assault held sway for much of the first half of 1942, eyes on both sides of the Atlantic (albeit reluctantly in some cases) increasingly began to visualize multinational operations in the Mediterranean theater on a significant scale. Ultimately, it was a combination of political design, strategic opportunity, and crisis as well as established in-theater momentum that led to the dispatch of the first USAAF units to the Middle East in June 1942.

The first significant contemplation of American air power being deployed to the Mediterranean theater in Allied strategic deliberations after Pearl Harbor occurred at the Arcadia Conference, held in Washington, DC, in December 1941 and January 1942. British prime minister Winston Churchill in particular was enthusiastic about the opportunities created for the employment of Allied air power by the new status of the United States as an active belligerent. In a lengthy memorandum on future strategy presented to Roosevelt in early January 1942, Churchill thought that "American squadrons, based primarily on Egypt and extending westwards, would be invaluable." This was even more so while the Axis air forces in the theater remained a powerful threat, and RAF resources were largely committed to operations against the Axis threat in Europe: "It would be to our advantage to develop air war in the Mediterranean on the largest scale on both sides, with constant bombing of enemy airfields and sea-traffic. . . . In the Mediterranean the enemy shows an inclination to develop a front, and we should meet him there with the superior strength which the arrival of the American Air Forces can alone give."[18]

Several days later, in light of possible diversions to other theaters, the British Chiefs of Staff pressed their American opposites to send "one or more American heavy bomber groups to Egypt" as the "most efficient method of improving Middle East situation."[19] In doing so, they were following the lead established by senior theater commanders, who made an early plea for USAAF units to be employed in the region. Churchill and the British Chiefs of Staff were advised that "if the American effort is to be directed into the right channel," then "the Mediterranean theatre offers good, if not the best, strategical prospects." Moreover, they continued, "facilities for the deployment of American Air Forces already exist in the Middle East and they could be used at once to great advantage."[20] Plans were also formalized at Arcadia for the future landing of an Anglo-American expeditionary force in French-held Northwest Africa, further indicating the potential of the wider Mediterranean within embryonic Allied strategic options.

Subsequent discussions and strategic alternatives contemplated by Allied leaders and their staffs between April and July 1942 were of particular importance to RAF Middle East for a number of reasons. First, they highlighted evolving Anglo-American political and military priorities and the considerations that shaped them. Second, they exerted a further positive influence on the prior momentum favoring direct American military intervention in the wider Mediterranean. Third, they had significant implications for the allocation of American-built aircraft to active combat theaters, including the Middle East.

The ultimate decision to commit to an Allied invasion of Northwest Africa in late 1942 (Operation Torch) was driven primarily by political considerations as well as by military and balance-of-power realities within the British-American partnership. Roosevelt and Churchill were equally determined that something had to be done in 1942, particularly in order to help their Soviet ally, even if what was done had little immediate impact on eastern front developments.[21] They were similarly united in their view that if the principle of "Germany first" was to be upheld, then Anglo-American forces must initiate offensive operations at a point that offered the best chances of success, especially for America's largely untried and inexperienced units.[22] Concurrently, Roosevelt was also open to the possibility of "direct aid to the Middle East in the shape of land and air forces" and the dispatch of USAAF units to Russia "to be maintained by way of the Persian Gulf."[23] This perspective would play a key role in the eventual dispatch of the first American combat units to the Middle East at a critical juncture in the desert fighting in mid-1942. And although these detachments were intended to be only short-term, the introduction of another guiding principle—that forces should be employed in those regions best prepared to receive and operate them— meant that their presence in the Middle East would very quickly become permanent.

The Issue of Aircraft Allocation and Its Wider Implications

Where potential combined air operations in the Middle East were concerned, perhaps the most immediate impact of the prolonged Anglo-American debate over strategic direction and the ultimate decision to implement Torch involved the continued allocation of aircraft to the theater. Here the growing possibility of American air units being committed to supporting British operations in the Mediterranean, though seemingly beneficial, presented a conundrum for RAF Middle East. A dynamic tension existed as to how these forces were to be equipped and sustained and how these actions might affect continued British access to American aircraft production. For example, in March 1942 the United States considered sending out USAAF combat personnel to the Middle East, but an integral part of these proposals was that the British would furnish the aircraft as well as maintenance, supplies, replacement planes, and training from their own resources.[24] Thereafter, with the United States seeking decisive action in Europe at the earliest possible opportunity and its military expansion gathering momentum came a growing desire to create complete units comprising aircraft and crews that would operate as the USAAF in various theaters rather than continuing to supply Britain with American-built aircraft for its own use.[25] As a consequence, there was a heightening consternation among British officials both in Washington and London as to the implications of this development for their own forces, especially in theaters outside the United Kingdom.[26] It was necessary, Dill stressed to General George C. Marshall, the US Army chief of staff, to continue the diversions of American aircraft in order "to prevent our losing the war while we are preparing to win it." From a British perspective, he continued, "very serious consequences" would arise from any sudden alterations in the allocation of aircraft production: "We have staked our whole development on this flow of American types [of aircraft], and the effect of any sudden interruption would be felt in every theatre. In the Middle East our whole position depends upon American types, and a sudden loss of deliveries would be a disaster."[27]

From early 1941, the reliance of Tedder's command on a continued supply of American-built aircraft was clearly evident, and this dependence intensified through 1942. The need for long-range heavy bombers in the theater and opportunities for their provision from American sources became a recurring theme. Four Boeing B-17s were supplied to Tedder in November 1941 for testing purposes, but when they began to suffer technical problems after only three sorties, they were withdrawn from service.[28] By mid-December, impressed with the four Boeing B-24s that were being operated by RAF Middle East, Tedder expressed his regret that there were too few of them. Furthermore, current allocations were unlikely to provide the three squadrons (complete with reserves and spares) that he estimated were necessary to allow sustained operations over a two-month period.[29] In February 1942, the British Chiefs of Staff recognized that the basing of American heavy-bomber groups in Egypt "would be of considerable strategic value" against enemy supply lines as well as naval and oil targets.[30]

Further weight to these claims came from Sir Stafford Cripps, Lord Privy Seal, when he passed through Cairo in March 1942. Asked by Churchill to assess the strategic situation there while he was en route to India, Cripps was quick to identify RAF Middle East's operational deficiencies, especially "the important question of bomber power," asserting, "I am confident heavy bombers can be used to better purpose here than against Germany." As a result, Cripps recommended to London that the dispatch of heavy and light bombers (including those from the United States) be expedited by "every possible means."[31] Between April and June 1942, theater authorities repeatedly stressed the need to undertake sustained aerial attacks on Axis airfields on nearby Sicily as a means of assisting the air defense and passage of much-needed supply convoys to the beleaguered island of Malta.[32] During this period, Colonel Bonner Fellers, the American military attaché at Cairo, also cabled Washington about the necessity of employing heavy bombers in the Middle East.[33]

RAF Middle East's growing reliance on American-built tactical aircraft likewise became more pronounced throughout 1941–1942. Alongside the gradual escalation of Bomber Command's night offensive and the demands of the air defense of the United Kingdom, emerging efforts to create a tactical air force to support future Anglo-American operations in Occupied Europe assumed immediate priority in British strategic calculations.[34] The resultant importance of American production lines to support British operations elsewhere can be illustrated by the evolving composition of RAF Middle East's air strength. On 15 June 1941, Tedder had available a total of nine fighter squadrons, including two made up of Curtiss P-40C Tomahawks, and nine light-bomber squadrons, incorporating three made up of Martin Marylands. On the eve of the Crusader offensive on 18 November 1941, five of his thirteen fighter squadrons were flying Tomahawks, four of his eleven light-bomber units were operating Marylands, and one former Maryland squadron had converted to the newer Douglas Boston III.[35] By the end of 1941, it was not uncommon for sorties to be undertaken by formations composed solely of American-built aircraft.[36] This dynamic only intensified once better types of aircraft, such as the Curtiss P-40D Kittyhawk and the Martin A-30 Baltimore, began to enter service in 1942. When the British Chiefs of Staff reviewed the Middle East strategy in April 1942, they concluded that "no matter what situation develops in the Middle East, strong air forces there are essential." But the primary if not sole hope of achieving this "lay in the reinforcement of our resources by the American air force, especially fighters and light bombers."[37]

The Torch decision changed the dynamics of American assistance to Britain in the theater in some key respects. The increasing likelihood of the commitment of US military forces had short-term and long-term impacts, both positive and negative, on RAF Middle East. More immediate shortfalls in the supply of American-built aircraft had to be weighed against the eventual dispatch of active US combat units. That said, the momentum generated by the Torch deliberations also helped safeguard continued British access to American aircraft production. The logistical balancing act faced in resolving immediate British concerns while facilitating the future expansion of the USAAF across multiple theaters was ultimately addressed by the Arnold-Towers-Portal Agreement, signed on 21 June 1942. The RAF was to receive approximately 3,000 aircraft from American stocks in the second half of 1942, a total far less than the 5,900 originally specified in January 1942 but much healthier than the 1,000 initially proposed by Arnold.[38] The agreement also allowed the British to shape future deployment to a significant degree, specifying the dispatch of USAAF heavy-bomber, medium-bomber, and fighter groups to the Middle East, with the first elements scheduled to commence operations by September 1942.[39]

INITIAL AMERICAN DEPLOYMENTS AND THEIR COORDINATION

By late June 1942, the situation confronting Middle East Command had become critical. A resurgent Axis had overwhelmed the prepared positions of the 8th Army at Gazala, swiftly encircled and captured the fortress of Tobruk, and was threatening to advance to Cairo. The first American-operated heavy bombers appeared in the Middle East as the Gazala battles raged, although their numbers and operational orders precluded intervention in the desert fighting. The employment of the Halverson Provisional Detachment (HAL-PRO) under Colonel Harry Halverson in early June 1942 was based on a mixture of design, opportunity, and crisis. Taking these elements in order, the arrival of HALPRO represented an acceleration of the allocation program outlined in the Arnold-Towers-Portal Agreement. The element of opportunity reflected the swelling chorus of observer reports not only outlining the invaluable addition that heavy bombers constituted but also highlighting the capacity of theater facilities to maintain them and various targets against which their employment would be most profitable. Indeed, the original rationale behind Halverson's detour to the Middle East was a surprise attack on the Ploesti oilfields, an all-too-familiar target, encouraged by Tedder, Bullitt, and others. The element of crisis was also in part a legacy of Anglo-American opinion as to the essential need for such aircraft in both defending Malta and acting as an alternative to the island's striking forces. After the raid on Ploesti and following repeated British approaches to the War Department, Halverson was given latitude to assist the RAF effort to cover the eastward leg of the Operations Vigorous and Harpoon relief convoy to Malta and to attack enemy ports.[40]

HALPRO was ultimately amalgamated with a small bomber force under Major General Lewis Brereton that arrived in late June to form the 1st Provisional Bomb Group. With the survival of both Egypt and Malta under grave threat, Brereton assumed control of the newly activated US Army Middle East Air Force (USAMEAF).[41] The American military hierarchy's decision to expedite the dispatch of additional heavy bombers to the Middle East was an "emergency" measure reflecting both Allied solidarity and the wider American investment at stake. The region held increasing strategic importance as an interdependent supply hub connecting the United States with both the Russian and the China–Burma–India theaters.[42]

Despite the sense of crisis that permeated the Middle East theater, several favorable aspects not only warranted but also expedited the dispatch of USAAF combat forces there. Since 1941, an extensive and increasingly flexible system of air supply and maintenance had been established in the region, chiefly through the combined efforts of British and American military and civilian authorities. For example, Halverson's force swiftly traversed the West African air route to reach the Middle East and was able to commence operations shortly after arriving due to the facilities and assistance provided by the British in Egypt.[43] Brereton's initial detachments similarly went into action within days after their arrival from India. The fighter units that embarked for the Middle East as part of USAMEAF were delivered successfully via the trans-African route, despite the fact that this avenue had been largely restricted to twin-engine aircraft since mid-1941.[44] In July 1942, the P-40 Warhawks of the 57th Fighter Group were ferried by the aircraft carrier USS *Ranger* to a point off Accra on the West African coast, from where they flew the well-traveled air route across the continent to Cairo.[45] Later that month, the 12th Medium Bombardment Group flew its B-25s out to Egypt via the recently opened southern route, using Ascension Island as a refueling point between Brazil and Africa.[46] Brereton was also able to use the established headquarters of Maxwell's Military Mission (redesignated US Army Forces in the Middle East) and the experience of members of the mission's former Air Section to coordinate operations with the British. For one of these members, Elmer Adler, it must have been a familiar feeling as he took command of the embryonic Air Service Command, USAMEAF. Although USAAF air-servicing personnel invariably arrived in the theater by ship and thus sometime after the actual combat units arrived, help was at hand not only from RAF Middle East but also from representatives of US aircraft and engine manufacturers based at the Gura and Abadan maintenance and repair depots.[47] In time, this growing military, logistics, and administrative presence demanded a further organizational change to reflect its status and permanence, and in November 1942 the 9th Army Air Force was established.

In other respects, the USAAF units had arrived at an opportune time. During June and July 1942, the RAF had reached a pinnacle of operational activity as it assisted the 8th Army in the Gazala battles and then successfully covered the army's retreat from Cyrenaica to Egypt. On repeated occasions, Air Vice Marshal Arthur Coningham's (head of the WDAF) fighters and bombers had intervened to slow the Axis advance and spare the retreating troops, trucks, and armor further serious losses from enemy air attack. Hard-won air superiority enabled the retiring 8th Army columns strung out along the desert coast road to remain in being.[48] The inspirational nature of the achievement to Anglo-American observers is neatly encapsulated in the RAF official account:

> In this period of extreme danger, in which the whole structure of British power to continue waging the war in the Middle East was threatened with collapse, the most reassuring factor was the way in which the RAF, far from being weakened by the enemy success on land, seemed to gain an added strength by a retreat on to their main bases and the resultant shortening of their lines of communication. . . . The RAF, in fact, far from being a beaten force[,] was intact, with morale particularly high, and secure in the knowledge that, in spite of the disaster which had overtaken the ground forces, it still had the measure of the enemy air forces.[49]

Richard Casey, the British minister of state in Cairo, informed Churchill that Tedder was "a tower of strength and common sense" during this critical period.[50] Little wonder, then, that General Sir Claude Auchinleck, commander in chief Middle East, thought afterward of "our air superiority" as "a very considerable, if somewhat indefinable asset."[51]

American military observers in the region drew similar conclusions. Brereton thought that "the RAF is due a major share of the credit for stopping Rommel's forces on the El Alamein line" and had "undoubtedly saved the Eighth Army from complete defeat."[52] Major Henry Cabot Lodge of the US Army Tank Corps thought that the theater was "fundamentally an air theatre—even more of an air theater than an armament theater." Whereas he thought little of the British Army in comparison to the Afrika Korps—calling the former "good amateurs against an army of professionals"—Lodge was more complimentary of the air effort. "The impression I got of the RAF," he informed the War Department on his return, "was of a more live, peppy, realistic organization than the British Army."[53]

Entering this atmosphere at such a time, the USAAF could not help but be impressed by the achievements of RAF Middle East, in particular its tactical air component. They were also provided with an accelerated practical education regarding air power's impact on maneuver warfare. Tedder, Coningham, and their respective staffs had given a potent demonstration of what a modern air force could achieve in trying circumstances. Clearly displayed were the basic tenets that underlay the modern application of air power that Tedder had been seeking to put into practice since mid-1941: air superiority as a vital prerequisite to other air, land, and sea operations; flexibility through concentration of force under a single air commander; and the need to recognize the interdependence of arms. The latter was especially important where the Middle East and the Mediterranean were concerned as campaigns in these theaters invariably became a "battle for airfields," with the victor enjoying significant advantages. As Tedder commented after the war, "It was no use having a victorious and predominant surface fleet if it was not free to operate because we had lost control in the air; it was no use having a strong army if, for the same reason, it could not be supplied and maintained."[54] In October 1942, Churchill was to give the ultimate vote of confidence, declaring: "Whenever our Army is established on land, and is conducting operations against the enemy, the system of organisation and employment of the Royal Air Force should conform to that which has proved so successful in the Western Desert."[55]

The WDAF's versatility and flexibility in joint operations was further exemplified by its multifunctional capability. This was undoubtedly impressed upon USAMEAF observers through the widespread employment of American-built Kittyhawks as fighter-bombers, a role relatively new to them.[56] The advantages of fighter-bombers armed with a 500-pound payload were many. By virtue of their speed, they could provide more responsive close support than conventional light bombers. Because of their rugged construction and ability to absorb punishment, they were able to operate at low levels over a confused battlefield and pick out enemy targets. Once their bombs were released, they were able to act as fighters, which in turn placed a reduced burden on maintenance facilities. They were also able to collect tactical intelligence to supplement other sources. In many respects, the fighter-bomber had rendered the dive bomber obsolescent and highlighted the latter's limitations when it was operating without air superiority or against heavily defended targets.[57]

Equally impressive from an American perspective were the efficient maintenance and repair systems and all-round mobility displayed by the WDAF. The progressive refinement of measures introduced in 1941 to allow continuous operations was paying huge dividends. It was not uncommon for individual squadrons to average more than 20 sorties a day for an entire month and for the WDAF's daily sortie rate to surpass 300 as operational records were repeatedly surpassed. And despite this extremely intensive workload, the WDAF's overall serviceability rarely dropped below 75 percent and actually increased during periods of battle.[58] By organizing fighters into wings and splitting individual squadron servicing crews into advance and rear parties, units were able to "leapfrog" each other, whether advancing or withdrawing, with little decline in efficiency. Air formations "were now interchangeable," and "squadrons could now be farmed out instantaneously to temporary foster-parents."[59] Perhaps most instructive was that all this was once again being done with ever-larger numbers of American-built aircraft. The WDAF order of battle on 26 May 1942, for example, contained a total of seven squadrons of American-built fighters (six of Kittyhawks and one of Tomahawks), compared to five squadrons of British-built fighters (four of Hurricanes and one of Spitfires). Moreover, the light-bomber complement was composed exclusively of American aircraft types—two squadrons of Bostons and one of Baltimores.[60]

Some aspects of RAF procedure were not totally new to USAAF personnel arriving in the theater, given the theory taught at the Air Corps Tactical School at Maxwell Field, Alabama.[61] What was revolutionary, though, was the opportunity to see these lessons translated into operational practice by experienced and trained pilots and crews. Brereton commented at length to Arnold about the efficient coordination of air-to-ground support and the command arrangements involved. Other USMEAF personnel interviewed on their return from the theater offered glowing praise of various aspects of the RAF setup. Adler thought the success achieved in close-support specifically and air operations more widely was due largely to the "mild-mannered" Tedder and "energetic" Coningham.[62] Lieutenant Colonel C. V. Whitney, assistant air intelligence officer, paid tribute to the sound application of the principles of air warfare and especially to the attention given to acquiring and using intelligence by RAF Middle East. In this arena, Tedder "had to have the information coming in constantly, right where he could see it."[63] At this juncture, the AOC-in-C was receiving strategic appreciations effectively straight from his opponents' headquarters, courtesy of Bletchley Park's Ultra operation, as well as intelligence from photographic and tactical-reconnaissance aircraft and important signals intelligence from Y intercept personnel.[64]

CHALLENGES TO GREATER INTEGRATION

The USAMEAF steadily grew over successive months to encompass multiple fighter-, medium-, and heavy-bomber units.[65] The British should have been overjoyed to see swelling numbers of available American aircraft, not just in terms of their impact over and beyond the immediate battlefield but also as a sign of further US commitment to the theater as a whole. Yet for Tedder and RAF Middle East, their appearance in some ways created more potential problems than it resolved outstanding ones. For example, most of the USAAF units had arrived with only minimal ground support staff and had to rely heavily on British resources until reinforced. Again, although highly trained as individuals, American pilots were relative novices in terms of the tactical cooperation methods developed and utilized by other air units and ground and naval forces.[66] As one commentator observed a few years after the war, "Peter would have to be robbed to pay Paul, but Paul was for the time being less experienced than Peter, and was therefore of less immediate value."[67] The major problem facing Tedder and Brereton, therefore, was how these forces were to be integrated into theater operations at a time when the ongoing campaign demanded that air power play a prime role in containing General Erwin Rommel's push to Cairo and laying the foundations for a successful 8th Army counteroffensive.

Command and control of national air contingents was also a potentially thorny issue. The American viewpoint before entering the war strongly advocated an organizational structure that upheld national and service independence. Brett, for example, stressed the strict supervision of USAAF units via a solely American, operational headquarters. Here he drew an explicit parallel with the American Expeditionary Forces sent to France in 1917 under General John J. Pershing. He was accordingly determined that any future dispatch of actual combat units to the theater should be based on the assumption that they would operate as a distinct, separate force, in effect a coequal of the RAF in all matters. "With all being under American control," he insisted, "there should be a definite and separate allocation of port facilities, supply lines and separate airdromes." Most importantly, "a definite understanding should be had as to the employment as an expeditionary force, and not as part of the Royal Air Force."[68]

This emphasis on distinct forces in turn highlighted another early point of doctrinal difference (and potential tension)—namely, air–land partnerships and the provision of air support on the battlefield. This difference was not entirely surprising, given that the USAAF remained an integral component of the US Army, unlike the independent status enjoyed by the RAF among the other British services. Where air support of ground forces was concerned, Brett concluded bluntly, the British Army leadership had "one big complaint"—they had to ask the RAF's permission for the use of air power.[69] Brigadier General Raymond Lee, the US military attaché in London, expressed similar misgivings and misapprehensions about the control of Middle East air resources. He cabled the War Department in late September 1941 that "near-sighted RAF officials" in the theater had "previously hampered all efforts to solve this long-standing problem of air support for ground forces." "There must be a plan of organization," he concluded, "superior to one based on good fellowship."[70] Even as late as September 1942, Major G. G. Atkinson, who had been the assistant military attaché in Cairo, was reiterating these criticisms. Atkinson remarked that the British policy of having a separate air force and a separate ground force (although fine in theory) was counterproductive to the provision of effective close air support in the field and provided a strong case against the future creation of an independent American air arm.[71] In the event, however, prolonged exposure to an environment that required harmonious multinational and interservice cooperation and the demonstrated effectiveness of the 8th Army–WDAF partnership would challenge these perceptions and allay these concerns.

ESTABLISHING A SUCCESSFUL FORMULA AND SUSTAINING MULTINATIONAL PARTNERSHIPS

The growing success achieved by the combined Anglo-American air effort in the Western Desert from August 1942 ironically owed much to the very factors that many American observers had reacted negatively against during 1941: improvisation, a flexible interpretation of doctrines, and a pooling of resources and facilities. Rather than operating independently, from the outset American fighter and bomber crews were fed, plane by plane and formation by formation, into existing WDAF wings and groups. This process usually commenced with squadron commanders and flight leaders and continued until the entire unit had become familiar with the realities of desert air warfare.[72] There inevitably were growing pains and teething troubles. Partway through this infiltration process, Tedder admitted that things were far from perfect, but he thought that the Americans were making pleasing progress, considering "there are a lot of things to be learned in this type of warfare, and they are not taught either in our schools or in your training in the States."[73] In the Western Desert, they progressively accumulated the training, techniques, and tactics necessary to conduct air warfare as part of a larger interdependent air arm. The strict delineation of operational command and control in the manuals often became somewhat blurred, but, most importantly, neither efficiency nor identity was sacrificed because after a brief period of indoctrination USAAF personnel resumed duties as all-American units that were part of the WDAF. Even then, though, they shared refueling and maintenance facilities with their British, South African, and Australian counterparts on a regular basis.[74]

A key component underlying these developments was the cordial working relationships necessary at various levels of the Anglo-American military hierarchy. Sir Charles Portal, Chief of the Air Staff, advised Tedder in mid-1942, "I hope you will go easy in your differences with the Americans, which are of course inevitable seeing that they cannot admit the inferiority which they must feel. No effort on your part could be too great to ensure that the two Air Forces do not get up against each other."[75] Nevertheless, irrespective of the strategic complexity conditioning the release of American-built aircraft to the Middle East, the obvious operational reliance upon these aircraft ensured that continued good British relations with the American logistic and supply staff in the theater remained a priority well before USAAF combat units arrived. The rapport established with Brett, Bullitt, and Adler provided substantial foundations on which to build an expanded American air presence. In some ways, the arrival of Halverson and then of Brereton could be seen as a suitable reward for mutual awareness and diligence in this area.

Impetus for greater cooperation had also been provided by the altered status of the United States as an active belligerent. Tedder in particular noted a dramatic change. Whereas prior to Pearl Harbor, there appeared to be an atmosphere of supreme self-satisfaction among American personnel, this was quickly replaced by a rather chastened approach to war, where they were "not quite so convinced now that they knew all about everything."[76] Yet although the Americans, now firmly in the "real" war, were increasingly receptive as a junior partner in the Middle East, there was still a need to play close attention to the dynamics of multinational relations. In Washington, Dill noted in March 1942 that there were "very few people on this side of the Atlantic who have the smallest conception of the magnitude of the war in which they are involved, or the utter unpreparedness of America to cope with the situation." Moreover, although "a great many Americans realize that things are not going well," it seemed "most of them blame us [the British] for that."[77] He later asserted to Churchill that "the real trouble is that some of them [the Americans] are quite convinced that they can run before they can walk."[78] In light of this, the importance of appointing those who could "get on" to positions of combined responsibility became paramount.

Of particular significance was the establishment of good relations between Tedder and Brereton. As the British had already realized and the Americans were now realizing, in a theater where interdependence rather than independence of arms was essential, securing the right blend of personalities and functioning as a team rather than as individuals were equally vital. A certain amount of damage control may also have been required after Tedder's experiences with Halverson. Despite the RAF extending assistance and facilities to the HALPRO force at its base at Fayid, Tedder found Halverson "most unwilling to accept any operational advice or instruction" on the day prior to attacking Ploesti. Moreover, little heed appeared to have been paid to operational procedure at various stages of HALPRO's mission, with the result that Halverson's bombers became widely dispersed on their return journey. Tedder's conclusions regarding the episode were damning. "This fiasco," he informed the British Air Ministry, "confirms our first impressions that the standard of training of the crews in this force is such as to render them practically useless for any purpose." He also held out little hope for plans to employ them against the Italian fleet, though he insisted "that they carry Royal Navy observers."[79] In future, the USAAF heavy-bomber units would have much closer contact with RAF Middle East's own 205 Group through the provision of liaison staffs and advanced headquarters.[80]

Both Tedder and Brereton made compromises and employed a mixture of tact, diplomacy, and forthrightness to achieve larger Allied goals and objectives. For example, in September 1942 Tedder observed that although the USAAF fighter- and light-bomber units were "fitting in very well" with the WDAF, the same could not be said of the USAAF heavy-bomber squadrons, who "still had much to learn." Tedder broached his concerns with Brereton, who promptly went to "give them a complete shaking up." The AOC-in-C also appreciated his counterpart's efforts in forming a small operational headquarters to exercise greater control over the heavy bombers. This headquarters was situated in a building alongside Tedder's own headquarters, a move that Tedder felt was a "big step in the right direction," safeguarding "any tendency to try and form a separate American Air Headquarters."[81] Further evidence of the successful partnership was shown in Brereton's opposition concerning proposals to transfer elements of his command back to the China–Burma–India theater.[82] It was also demonstrated in the flexible arrangements subsequently established for the theater's long-range air strike components, with RAF Middle East's two heavy-bomber squadrons placed under the operational control of an American headquarters—Bomber Command, USMEAF—activated on 12 October.[83]

Nowhere were the results of this unified approach more evident than on the eve of the El Alamein offensive, 22 October 1942, when both Tedder and Brereton wrote lengthy cables to their respective air chiefs. Tedder informed Portal that "the Americans work in very well with our squadrons. . . . They are learning from us and we are learning from them—I was glad to hear this from both sides."[84] Of the thirty-seven squadrons assembled for the offensive, seven belonged to the USAAF as part of the newly formed American Desert Air Task Force. Furthermore, the organization of the WDAF involved USAAF, RAF, Royal Australian Air Force, and South African Air Force squadrons, all operating in fighter- and light-bomber wings as part of a synchronized whole.[85] Meanwhile, Brereton announced to Arnold that his squadrons "were ready for anything." Brereton himself led an advanced American air headquarters attached to the Advanced Headquarters of the WDAF, using "the Intelligence facilities of the RAF, British Army and Royal Navy, which have been made fully available to me."[86] Brereton's chief of staff, Colonel Auby Strickland, was well familiarized with WDAF practice and procedure, having been attached to Coningham's headquarters in late July 1942.[87]

The aerial campaign waged before, during, and after the El Alamein offensive confirmed these positive impressions. British and American squadrons operated in unison in the various air missions essential to achieving the 8th Army's designs: attaining air superiority, isolating the battlefield, and providing close support. It was increasingly common for separate WDAF and Desert Air Task Force formations not only to cooperate closely but also to do so exclusively in American-built planes. As the official Royal Australian Air Force account later noted, by late 1942 a "standard bombing force" consisted of "eighteen Baltimores or Mitchells led by six Kittyhawks with bombs and six without, while twelve Kittyhawks provided top cover, flying above and slightly behind the main force."[88] By early November, after a prolonged slugging match, Rommel had no option but to start withdrawing westward. That his battered forces were increasingly short of air support, critical supplies, and sleep was in no small way due to the employment of an Anglo-American air force, flying increasing numbers of American-built aircraft. Indeed, this outcome represented a remarkable progression from the multinational challenges confronting combined air operations in the theater only months earlier, and it was fitting that both American and British air units shared in the success achieved.

Conclusion

The creation of an Anglo-American air component in the Western Desert reflected a combination of complex and fortunate contextual factors. Powerful legacy elements, strategic reorientation, political and operational imperatives, and growing economic reliance played a part in generating favorable momentum. But in many respects these factors represented only part of the process. Although they clarify the genesis and early implementation of multinational air power in the theater, the explanation for its successful, enduring practical application lies elsewhere. If there was any identifiable "recipe" for success where the USAAF and RAF Middle East were concerned, several key ingredients stood out. Central was the role of personalities and how balancing tact and compromise with straight talking often provided a stronger adhesive and influence than rigid attachment to doctrine. Another feature was the bedrock of common objectives, against which the inevitable variety of challenging issues tended to founder. Yet another element reflected the growing recognition on both sides of their respective roles, alongside an awareness of their relative experience, strengths, and limitations. These factors ultimately helped to ensure the fashioning and functioning of an effective military coalition in terms of mutual understanding, shared goals, and effective execution on the battlefield.

Notes

1. Major General George Brett to General Henry Arnold, 17 September 1941, Henry Arnold Papers, box 41, Manuscript Division, Library of Congress (LoC), Washington, DC.

2. Colonel Gerald Brower to Arnold, 20 November 1940, Gerald Brower Papers, 168.6003, Air Force Historical Research Agency (AFHRA), Maxwell Air Force Base, AL.

3. Brower to Arnold, 23 December 1940, Brower Papers, 168.6003, AFHRA.

4. Brower to Arnold, 15 January 1941, Brower Papers, 168.6003, AFHRA.

5. Brett to Arnold, 5 October 1941, Elmer Adler Papers, 168.05-1, AFHRA.

6. Brett to Arnold, 29 September 1941, Adler Papers, 168.05-1, AFHRA.

7. Arnold to Brett, 3 October 1941, and Brett to US War Department, 24 October 1941, Adler Papers, 168.05-1, AFHRA.

8. Brett to Arnold, 15 and 24 September 1941, Adler Papers, 168.05-1, AFHRA.

9. William Bullitt to Franklin D. Roosevelt, 20 December 1941, in US Department of State, *Foreign Relations of the United States: The Conferences at Washington, 1941–1942, and Casablanca, 1943* (Washington, DC: US Government Printing Office, 1968), 47–49.

10. Bullitt to Roosevelt, 21 December 1941, in US Department of State, *Foreign Relations of the United States,* 49–50.

11. Field Marshal Sir John Dill to Winston Churchill, 7 March 1942, PREM 3/478/6, The National Archives of the United Kingdom (TNA), London.

12. Lieutenant Colonel Elmer Adler to Brett, 23 December 1941, Adler Papers, 168.05-4, AFHRA.

13. Adler to Muir Fairchild, 3 January 1942, Adler Papers, 168.05-4, AFHRA.

14. Adler to US War Department, 10 December 1941, Adler Papers, 168.05-15, AFHRA.

15. Adler to Brett, 23 December 1941, Adler Papers, 168.05-4, AFHRA; Adler to Arnold, 22 January 1942, Adler Papers, 168.05-14, AFHRA.

16. Adler to Arnold, 4 February 1942, Adler Papers, 168.05-15, AFHRA.

17. Adler to Brett, 13 February 1942, Adler Papers, 168.05-4, AFHRA. For a summary of American air logistics assistance to the theater before 1942, see Christopher M. Rein, *The North African Campaign: US Army Air Forces from El Alamein to Cassino* (Lawrence: University Press of Kansas, 2012), 41–42.

18. Churchill to Roosevelt, c. 7 December 1941, C-153x, in *Churchill & Roosevelt: The Complete Correspondence,* vol. 1, ed. W. F. Kimball (Princeton, NJ: Princeton University Press, 1984), 317–18.

19. Chiefs of Staff Committee to Chiefs of Staff, 10 January 1942, PREM 3/439/10, TNA.

20. Middle East Defence Committee to Churchill and Chiefs of Staff, 27 December 1941, PREM 3/310/3, TNA.

21. War Cabinet, minutes of a meeting held at 10 Downing Street, 22 May 1942, PREM 3/333/8, TNA; War Cabinet, minutes of a meeting with the Soviet Delegation held at 10 Downing Street, 9 and 15 June 1942, WP (42) 254, PREM 3/333/8, TNA; Dill to Chiefs of Staff, 15 June 1942, PREM 3/476/11A, TNA.

22. Louis Mountbatten to Churchill, 16 June 1942, PREM 3/476/11A, TNA; Dill to Chiefs of Staff, 31 July 1942, PREM 3/439/11, TNA; Churchill to Dill, 31 July 1942, PREM 3/439/11, TNA; Dill to Churchill, 1 August 1942, PREM 3/439/11, TNA; Chiefs of Staff to Churchill, 7 August 1942, PREM 3/439/13, TNA.

23. Dill to Churchill, 15 June 1942, PREM 3/476/11A, TNA.

24. Joint Staff Mission to Chiefs of Staff, 19 March 1941, PREM 3/492/1, TNA.

25. Arnold to Sir Charles Portal, 30 May 1942, PREM 3/453/2, TNA.

26. Joint Staff Mission to Chiefs of Staff, 5 March and 1 April 1942, PREM 3/11/12, TNA. See also Robert S. Ehlers, *The Mediterranean Air War: Airpower and Allied Victory in World War II* (Lawrence: University Press of Kansas, 2015), 193–94.

27. Dill to General George C. Marshall, 16 April 1942, PREM 3/453/5, TNA.

28. Portal to Churchill, 21 October 1941, PREM 3/291/7, TNA.

29. Air Marshal Arthur Tedder to Wilfrid Freeman, 16 December 1941, AIR 23/1386, TNA.

30. Chiefs of Staff to Joint Staff Mission, 23 February 1942, PREM 3/492/1, TNA.

31. Stafford Cripps to Churchill, 20 March 1942, PREM 3/291/6, TNA.

32. William Dobbie to Chiefs of Staff, 30 April 1942, PREM 3/11/12, TNA; Churchill to Portal, 1 May 1942, PREM 3/11/12, TNA; Commanders in Chief to Chiefs of Staff, 25 June 1942, PREM 3/292/4, TNA.

33. Thomas Mayock, "The North African Campaigns," in *Europe: Torch to Pointblank,* vol. 2 of *The Army Air Forces in World War II,* ed. Wesley Frank Craven and James Lee Cate (Washington, DC: Office of Air Force History, 1983), 9, 15.

34. Winston Churchill, "A Review of the War Position," memorandum, 21 July 1942, WP (42) 311, PREM 3/499/3, TNA.

35. Air Historical Branch (AHB) Narrative, *Operations in Libya and the Western Desert, June 1941–January 1942,* vol. 2 of *The Middle East Campaigns,* 54–55, AIR 41/25, TNA.

36. Ibid., 2:164, 187, 231.

37. Chiefs of Staff to Joint Staff Mission, 1 April 1942, PREM 3/492/2, TNA.

38. John Slessor to Churchill, 18 June 1942, PREM 3/453/2, TNA; AHB Narrative, *Anglo-American Collaboration in the Air War over NW Europe 1940–1942,* 137–38, 145–48, 154–56, AIR 41/62, TNA.

39. Memorandum of Agreement between Lieutenant-General Arnold, Rear-Admiral Towers, and Air Chief Marshal Portal, 21 June 1942, PREM 3/453/2, TNA. See also Gavin J. Bailey, *The Arsenal of Democracy: Aircraft Supply and the Anglo-American Alliance, 1938–1942* (Edinburgh: Edinburgh University Press, 2013), 199.

40. Chiefs of Staff to Joint Staff Mission, 5 June 1942, PREM 3/492/2, TNA; Dill to Chiefs of Staff, 6 June 1942, PREM 3/492/2, TNA; Chiefs of Staff to Dill, 7 June 1942, PREM 3/492/2, TNA; Dill to Chiefs of Staff, 10 June 1942, PREM 3/492/2, TNA.

41. USAAF Historical Division, *The AAF in the Middle East: A Study of the Origins of the Ninth Air Force,* Study no. 108, p. 68, USAF Collection, 101.108, AFHRA.

42. Dill to Chiefs of Staff, 1 July 1942, PREM 3/492/2, TNA.

43. John Carter, "The Early Development of Air Transport and Ferrying," in *Plans and Early Operations, January 1939 to August 1942*, vol. 1 of *The Army Air Forces in World War II,* ed. Wesley Frank Craven and James Lee Cate (Washington, DC: Office of Air Force History, 1983), 341–42.

44. AHB Narrative, *The West African Air Reinforcement Route,* vol. 10 of *The Middle East Campaigns,* 102–4, AIR 41/32, TNA.

45. USAAF Historical Division, *The AAF in the Middle East,* 75–76.

46. Ibid., 79–80.

47. Ibid., 67, 69, 81–84.

48. AHB Narrative, *Operations in Libya and the Western Desert (Including Malta),* vol. 3 of *The Middle East Campaigns,* 21 January–30 June 1942, 208, 223, AIR 41/26, TNA.

49. AHB Narrative, *Operations in Libya, the Western Desert, and Tunisia, July 1942–May 1943,* vol. 4 of *The Middle East Campaigns,* 81–82, AIR 41/50, TNA.

50. Richard Casey to Churchill, 25 June 1942, PREM 3/292/6, TNA.

51. General Sir Claude Auchinleck, "Operations in the Middle East from 1 November 1941 to 15 August 1942," *Supplement to the London Gazette,* 13 January 1948, USAF Collection, 512.952A, AFHRA.

52. Lewis H. Brereton, *The Brereton Diaries: The War in the Air in the Pacific, Middle East, and Europe, 3 October 1941–8 May 1945* (New York: Morrow, 1946), 137.

53. Major Henry Cabot Lodge III, interview, 7 July 1942, USAF Collection, 142.052, AFHRA. See also US War Department, Military Intelligence Service, "Notes and Lessons on Operations in the Middle East," Campaign Study No. 5, 30 January 1943, USAF Collection, 170.223-5, AFHRA.

54. Sir Arthur Tedder, "Air, Land, and Sea Warfare," *Royal United Service Institution Journal* 91 (1946): 61–64.

55. Churchill to James Grigg and Archibald Sinclair, 7 October 1942, PREM 3/8, TNA.

56. Brereton, *The Brereton Diaries,* 143.

57. AHB Narrative, *Operations in Libya and the Western Desert (Including Malta),* 3:73, 98, 141–42, 149, 216–17; Brad William Gladman, *Intelligence and Anglo-American Air Support in World War Two: The Western Desert and Tunisia, 1940–43* (Basingstoke, UK: Palgrave Macmillan, 2009), 87.

58. AHB Narrative, *Operations in Libya and the Western Desert (Including Malta),* 3:179–82, 192, 220; John Herington, *Air War against Germany and Italy, 1939–1943* (Canberra: Australian War Memorial, 1962), 244–47.

59. Roderic Owen, *The Desert Air Force* (London: Hutchinson, 1948), 73.

60. AHB Narrative, *Operations in Libya and the Western Desert (Including Malta),* 3:appendix 17.

61. See late prewar Air Corps Tactical School lectures on aspects of air employment in USAF Collection, 168.7045-35, AFHRA.

62. Major General Lewis Brereton to Arnold, "Direct Air Support in the Libyan Desert," memorandum, 3 November 1942, USAF Collection, 145.96-64, AFHRA; Brigadier General Elmer E. Adler, interview, 18 January 1943, USAF Collection, 142.052, AFHRA; Colonel Robert G. Gard and Colonel Clifford G. Kershaw, interviews, 31 December 1942, USAF Collection, 142.052, AFHRA.

63. Lieutenant Colonel C. V. Whitney, interview, 6 April 1943, USAF Collection, 142.052, AFHRA.

64. Group Captain R. H. Humphreys, "The Use of 'U' in the Mediterranean and Northwest African Theaters of War," SRH-037, "Reports Received by U.S. War Department on Use of Ultra in the European Theater, World War II," USAF Collection, MICFILM 34090, AFHRA; Gladman, *Intelligence and Anglo-American Air Support in World War Two*, 90, 102–3.

65. Mayock, "The North African Campaigns," 2:11–14.

66. Roderic Owen, *Tedder* (London: Collins, 1952), 166–67.

67. Owen, *The Desert Air Force*, 120.

68. Brett to Arnold, 5 October 1941, Adler Papers, 168.05-4, AFHRA.

69. Brett to Arnold, 17 September 1941, Arnold Papers, box 4, Manuscript Division, LoC.

70. Brigadier General Raymond Lee to US War Department, 27 September 1941, Adler Papers, 168.05-1, AFHRA.

71. Major G. G. Atkinson, interview, 24 September 1942, USAF Collection, 142.052, AFHRA.

72. Colonel Frank Mears, interview, 10 August 1943, USAF Collection, 142.052, AFHRA; Major P. R. Chandler, interview, 17 June 1943, USAF Collection, 142.052, AFHRA; USAAF Historical Division, *The AAF in the Middle East*, 76–77, 97–98.

73. Quoted in Brigadier General John E. Upston, interview, 8 December 1942, USAF Collection, 142.052, AFHRA.

74. USAAF Historical Division, *The AAF in the Middle East*, 98–99; Mayock, "The North African Campaigns," 2:26–27, 33; Owen, *Desert Air Force*, 121; Owen, *Tedder*, 167.

75. Portal to Tedder, 21 July 1942, Viscount Portal Papers, box C, folder 8, item 21a, Lord Portal Collection, Christ Church College, Oxford.

76. Lord Tedder, *With Prejudice: The War Memoirs of Marshal of the Royal Air Force, Lord Tedder* (London: Cassell, 1966), 209–10.

77. Dill to Churchill, 7 March 1942, PREM 3/478/6, TNA.

78. Dill to Churchill, 9 May 1942, PREM 3/478/6, TNA.

79. Tedder to Douglas Evill, 14 June 1942, PREM 3/374/6, TNA.

80. USAAF Historical Division, *The AAF in the Middle East*, 75, 79.

81. Tedder to Portal, 9 September 1942, Portal Papers, box C, folder 8.

82. USAAF Historical Division, *The AAF in the Middle East*, 85–86.

83. Ibid., 94–95. See also Vincent Orange, "Getting Together: Tedder, Coningham, and the Americans in the Desert and Tunisia, 1940–43," in *Airpower and Ground Armies: Essays on the Evolution of Anglo-American Air Doctrine 1940–1943*, ed. Daniel R. Mortensen (Montgomery, AL: Air University Press, Maxwell Air Force Base, 1998), 4–5.

84. Tedder to Portal, 22 October 1942, AIR 23/1398, TNA.

85. AHB Narrative, *Operations in Libya, the Western Desert, and Tunisia*, 4:223–37, 229–32.

86. Brereton to Arnold, 22 October 1942, Hap Arnold/Murray Green Collection, series 8, box 104, folder 6, Official Correspondence, 7 October–30 November 1942, US Air Force Academy, CO.

87. Brereton, *The Brereton Diaries*, 142.

88. Herington, *Air War against Germany and Italy*, 370.

The Royal New Zealand Air Force and American Supply in the South Pacific during the Second World War

John Moremon

The Royal New Zealand Air Force (RNZAF) was the least developed of the British Empire's air forces at the start of the Second World War, not established as an independent air force until 1 April 1937.[1] New Zealand's air policy was intended to meet the dual requirements of national and imperial defense. The historian Ashley Jackson notes that air policy before the war and early in the war seemed directed more toward assisting Great Britain's Royal Air Force (RAF) "in every possible way, rather than maintain[ing] a significant independent air force."[2] One of its roles was to channel aircrew trainees to the RAF—a role emphasized when New Zealand entered into the wartime British Commonwealth Air Training Plan, or Empire Air Training Scheme. Nevertheless, the RNZAF also played a role in the defense of New Zealand and certain islands in the South Pacific. After the Pacific War erupted, it contributed to the American-commanded South Pacific Area (SOPAC), in which it in effect came of age within a multinational air power environment.

The RNZAF's contribution to the Pacific War is generally not widely known, even in its own country. The first of two notable studies of that contribution was Squadron Leader J. M. S. Ross's official history *Royal New Zealand Air Force*, published a decade after the war.[3] Ross was expected to conform with the main functions of official war history in New Zealand at the time, which were to produce a record of wartime activities that could provide guidance in the event of a future conflict and to deliver to the general public (including veterans) an account of their country's participation in the war and a written memorial to those who lost their lives.[4] Ross portrayed RNZAF operations as both an achievement in that the small air force emerged as a regional air power and something of an anomaly in that this air force modeled on the RAF and trained along British lines was heavily influenced by the Americans but then realigned with the RAF straight after the war.[5] Four decades later, Brian J. Hewson delivered a scholarly reassessment of the RNZAF in his PhD dissertation "Goliath's Apprentice: The Royal New Zealand Air Force and the United States in the Pacific War, 1941–1945." Hewson argues that the RNZAF's modest contribution to operations in the South Pacific was vitally important for national strategy and air power development. He suggests that it underwent "a valuable apprenticeship" under American tutelage.[6]

The South Pacific was a challenging geographical environment. Operations tended to be heavily joint in nature, requiring cooperation between air, land, and naval forces to strike at enemy bases and shipping, conduct amphibious landings, and engage in jungle warfare. The tropics, with its constant heat, glare, heavy rains, humidity, dense vegetation, and diseases, was considered as much an enemy as the opposing forces. The SOPAC was also unusual in that naval, land, and air forces were all under naval strategic command. A US Army Air Forces (USAAF) official history of this area during the war, edited by Wesley Frank Craven and James Lea Cate, notes that the primary influences on Allied air power were the tropical environment and the fact that "[t]he South Pacific was a Navy theater; admirals commanded both its air and surface forces, regardless of parent service of the units involved."[7] The RNZAF was the smallest, least-well-equipped, and only non-American air force in the SOPAC. One of the key challenges faced in multinational operations, particularly when there is an asymmetrical relationship between coalition partners, is interoperability.[8] Both Ross and Hewson note that a process of expansion and reequipment was required for the Americans to utilize the RNZAF. This chapter considers how the dominant service partner in the SOPAC came to influence the types and numbers of aircraft operated by the RNZAF to ensure interoperability. Although this influence enabled the New Zealanders to contribute to the air war, it ultimately also imposed a practical constraint on the use of the RNZAF in other areas of the Pacific War.

Before the Pacific War

New Zealand was a small country of approximately 1.6 million people during the Second World War. Before the war, successive governments subscribed to the "Singapore strategy," wherein Great Britain pledged to dispatch a Royal Navy fleet to the Far East should British Empire interests in the Asia-Pacific be militarily threatened. As the southernmost country in the western Pacific, New Zealand was geographically remote and never likely to be invaded. From the outset, New Zealanders understood that their air force (when formed) would be maritime focused. The country's first report on air power requirements in 1919 noted that its aircraft, when acquired, could assist naval forces in countering raids against shipping lanes and harbors.[9] Yet even raids were considered unlikely. The lack of a perceived threat and challenging economic conditions thus conspired to delay the formation of an independent New Zealand air force.[10] In 1923, the New Zealand Army formed an administrative branch titled the Permanent Air Force, which was responsible for air-related administration and recruitment, as well as a territorial (reservist) branch titled the New Zealand Air Force. The latter's squadrons were tasked mainly with maritime patrolling. At the end of 1935, with tensions mounting in Europe and the Asia-Pacific, a new government resolved to strengthen defenses and establish an air force. Group Captain Ralph Cochrane, RAF, was tasked with reporting on its role and equipment needs. He recommended that the air force include two squadrons equipped with bombers capable of intercepting warships before they reached New Zealand and deploying to islands in the South Pacific.[11]

Cochrane became the RNZAF's first Chief of Air Staff (CAS). His air force was allotted three principal roles: cooperation with land and naval forces for the defense of New Zealand; cooperation with naval forces for the protection of maritime trade; and training of aircrew and ground staff for the RNZAF and RAF.[12] In April 1939, a British–Australian–New Zealand defense conference pointed to the possibilities of a war against Germany or concurrent wars against Germany and Japan. Trans-Pacific and trans-Tasman trade was believed vulnerable in either event. The New Zealand government accepted that the RNZAF required aircraft capable of long-range patrolling and so ordered thirty Vickers Wellington bombers from Britain. The Wellingtons were not delivered before the outbreak of war, so they were offered to the RAF to equip a bomber squadron in Britain. The RNZAF continued operating

obsolescent biplanes, including ten Vickers Vildebeest torpedo bombers purchased in the mid-1930s and a couple-of-dozen secondhand Blackburn Baffin
naval torpedo bombers.

New Zealand's main contributions to the war in Europe were the 2nd
New Zealand Division, deployed to the Middle East in 1940, and participation in the British Commonwealth Air Training Plan. The country's home
defenses were built up in case of a war against Japan, but equipment and
training were never satisfactory. Britain supplied several dozen training aircraft as well as some eighty Vickers Vildebeests and Vincents—essentially
the same aircraft but with torpedo equipment removed and the option of
an auxiliary fuel tank in the Vincents—that had been retired from RAF service. With only lumbering biplanes for maritime patrols, the RNZAF had
little hope of detecting German raiders that ventured into the South Pacific
in 1940. In November that year, the RNZAF resorted to engaging Tasman
Empire Airways to search for raiders using Short Empire flying boats.[13]

In the same period, New Zealand sent a force to Fiji, which was deemed
vulnerable in the event Japan entered the war. Vildebeests and Vincents were
thought to be unsuitable for service in the tropical islands. The RNZAF dispatched a flight of De Havilland Dragon Rapides—biplane airliners impressed
into service and converted into navigation and light-bombing trainers—
to patrol the seas around Fiji.[14] Prime Minister Peter Fraser (1940–1949)
pleaded with his British counterpart, Winston Churchill (1940–1945, 1951–
1955), for more aircraft, but to no avail. Fraser informed Churchill that he
was "keenly disappointed," particularly given that New Zealand had willingly
offered up its Wellingtons to the RAF, "which, if only a few had been delivered
here, would have relieved us of our present very grave anxieties."[15] Fraser next
approached the Australian government for three Consolidated-Vultee PBY
Catalina flying boats, but the Australians responded that none could be
spared.[16] In essence, the RNZAF was a maritime patrol force that was lacking
maritime patrol capability.

EARLY COOPERATION WITH THE AMERICANS

In 1941, as war against Japan appeared likely, the British Empire, the United
States, and the Netherlands East Indies agreed to cooperate militarily. The
United States was obviously the dominant ally, but it did not appear to be
overly concerned with the South Pacific. The US Navy's War Plan Orange

anticipated fleet operations and a decisive naval battle in the mid-Pacific. However, the United States maintained a colonial presence in the South Pacific, having annexed the eastern Samoan Islands in 1899 and moving to annex several uninhabited islands (previously claimed by Britain) south of Hawaii in the 1930s. The purpose of these annexations was revealed in 1941 when the US War Department authorized the development of an aircraft ferry route in the South Pacific. This was a precautionary measure because the mid-Pacific ferry route used to deliver Boeing B-17 Fortress bombers to the Philippines was liable to be cut by the Japanese. The USAAF proposed to ferry B-17s (and other long-range aircraft) from the US west coast to Hawaii and then southwest to Australia, with refueling stops on five islands: the recently annexed Palmyra Island (later switched to Christmas Island), Canton Island, American Samoa, Fiji, and New Caledonia.[17] The New Zealand government welcomed this development and agreed to improve an airfield at Nandi, Fiji, to support B-17s.[18]

In September 1941, Air Commodore Victor Goddard, RAF, was appointed CAS of the RNZAF. A naval aviator during the First World War, he had recently served as a senior staff officer in the British Expeditionary Force in France and with the British War Office. The air force that Goddard assumed command of was ill equipped and undermanned but not in quite the same parlous state as a year earlier. Over the course of 1941, Britain supplied the Australian, Canadian, New Zealand, and South African air forces with Lockheed Hudson medium bombers to boost maritime patrol capabilities. At the start of the Pacific War, the RNZAF's first-line aircraft included thirty-six Hudsons, thirty-five Vincents, and four antiquated Short Singapore flying boats. Most were based in New Zealand, but several Vincents and the Singapores were in Fiji.[19] Fearing carrier-borne air raids, Fraser urged Churchill to send Hawker Hurricane fighters, again to no avail. Fortunately, another avenue of supply opened up in January 1942 when the British-American Combined Munitions Board was established in Washington. New Zealand's requests for equipment and munitions were then assessed by the British Chiefs of Staff (BCS) and forwarded to the Combined Munitions Board for determination. Following lobbying by Australia and New Zealand, the BCS and the board agreed these countries should receive further Hudsons and, critically, Curtiss P-40 Warhawk/Kittyhawk fighters. This decision required the BCS to divert aircraft intended for the RAF in the Middle East. The new aircraft were not expected in New Zealand until March–April 1942, and in the meantime the RNZAF made plans to use trainers in combat roles in the event of an emergency.[20]

After Japan's entry into the war, securing the South Pacific ferry route was a high priority for the Americans. Japanese victories made it apparent that the ferry route, which had been considered secure, could be threatened. US forces garrisoned Christmas Island, Canton Island, and American Samoa, while New Zealand boosted its force in Fiji, and Australia sent a modest force to New Caledonia to assist the Free French there. In Washington, DC, General Henry H. "Hap" Arnold, commanding general, USAAF, hoped the RNZAF and Royal Australian Air Force could provide aerial defenses for Fiji and New Caledonia, respectively, but neither possessed fighters. Arnold authorized the raising of Task Force "Five Islands" with a view to depositing a USAAF fighter squadron at each island base on the ferry route.[21]

American–New Zealand cooperation commenced in a modest fashion in January 1942 when US Navy (USN) Patrol Squadron 23 (VP-23) began patrolling between Canton Island and Fiji. Catalina crews based on Canton Island completed the 1,250-mile (2,012-kilometer) first leg of their patrol by landing at Suva, Fiji, where RNZAF personnel refueled the aircraft and hosted the crews before their return leg. That same month a Japanese submarine fired on a merchant cruiser off Fiji. In late January, the USAAF's 70th Pursuit Squadron, part of Task Force "Five Islands," arrived with Bell P-39 Airacobras. The Americans must have been surprised to find their Allies operating biplanes! The two general-reconnaissance squadrons based in Fiji, Numbers 4 and 5 Squadrons, were equipped with a mixture of Vincents and Singapores. Fortunately, the RNZAF had sent six Hudsons from New Zealand to be attached to Number 4 Squadron.[22] Group Captain Geoffrey Roberts, RNZAF, commanded Allied Air Forces in Fiji, but this was a temporary appointment because Americans would subsequently command all garrisons that contained American units. Other than an invasion scare in mid-February 1942, there was no action at Fiji.[23]

In the wake of Allied defeats in Malaya, the Philippines, the Netherland East Indies, and New Britain, attention turned to securing Burma and India in Southeast Asia and Australia, New Zealand, and those islands still in Allied hands in the South Pacific. The British-American Combined Chiefs of Staff established two strategic areas, with the BCS responsible for Southeast Asia and the US Joint Chiefs of Staff (JCS) responsible for the Pacific. In mid-April 1942, the JCS split the Pacific into the Southwest Pacific Area (SWPA), under

General Douglas MacArthur, US Army, based in Australia, and the Pacific Ocean Areas, under Hawaii-based Admiral Chester W. Nimitz, USN. The SOPAC, one of three areas making up the Pacific Ocean Areas, encompassed more than one million square miles (2.6 million square kilometers) of ocean dotted with thousands of islands and atolls. The JCS gave the USN command of all forces in the SOPAC except those reserved for New Zealand's defense, which remained under the New Zealand Chiefs of Staff (NZCS).[24]

Admiral Robert L. Ghormley, commander, SOPAC (COMSOPAC), USN, reached New Zealand in May 1942. Ghormley explained to the NZCS that he was not responsible for New Zealand's defense, and therefore he had no particular interest in the equipping and supplying of its forces. This was a blow for the RNZAF. Knockbacks from Britain whenever aircraft were requested had made it apparent that the RNZAF would become reliant on the United States for aircraft.[25] Without a theater commander prepared to advocate for it, the RNZAF was liable to be starved of aircraft and equipment. The New Zealand government lobbied the JCS to guarantee supply of the RNZAF and in the process proposed building up the air force to twenty squadrons. The JCS preferred a limit of ten squadrons—four light bomber, five fighter, and one army cooperation—which they deemed sufficient for New Zealand's defense needs. Determined that its forces would contribute to a counteroffensive, the New Zealand government continued pressing the JCS to reconsider the matter, taking the position that the RNZAF needed to be incorporated into the SOPAC because "the proper employment of air forces, particularly in support of naval operations, demands common doctrine and as far as possible complete operational coordination which can only be achieved under single direction."[26] The JCS finally relented in late 1942, at which time New Zealand also signed a "lend-lease" agreement with the United States to enable direct supply.[27]

Ghormley's first commander, Aircraft, South Pacific (COMAIRSOPAC), was Rear Admiral John S. McCain, USN. Having spent much of his career in battleships and cruisers, McCain retrained as an aviator at the age of fifty-two. He had commanded an aircraft carrier in 1937–1939 and had then been commander, Aircraft, Scouting Force, in the Atlantic Fleet in 1941. McCain reached New Caledonia in May 1942 and assumed command of Task Group 63, an air task force comprising USN, US Marine Corps (USMC), USAAF, and RNZAF land-based and flying-boat squadrons as well as ancillary units. The task force was widely dispersed across New Caledonia, Fiji, Tonga,

Samoa, and the New Hebrides, with airfield construction under way at all locations. USAAF officers mounted a campaign to ensure that units of each service would remain intact rather than be broken up and distributed across different islands.[28] McCain had to ensure that his squadrons could defend island bases and Allied shipping, patrol over the broad expanses of ocean and islands, and strike at enemy bases and shipping. With more than 1,000 miles (1,609 kilometers) between his most distant bases and difficulties in communications, McCain was not able to exercise command over all units directly. He outlined a general doctrine for the employment of his air force assets and made each service responsible for its own training and logistics, with the understanding that they would contribute to interservice air operations when required.[29]

To begin with, the New Zealanders played a modest role in Task Group 63. Maritime patrol increased in importance, however, as the Americans began assembling shipping for naval and amphibious operations in the Solomon Islands. Faced with a shortage of patrol aircraft in New Caledonia, McCain ordered the transfer of several Vincents from Fiji. He probably did not realize that they were biplanes. The RNZAF determined that Hudsons would be more suitable because they had greater range and were better armed. A detachment of two Hudsons arrived from Fiji in July 1942, and then more Hudsons, aircrews, and ground staff arrived from New Zealand to form a new general-reconnaissance squadron. The New Zealanders took over patrolling from American bomber crews, with their patrol area stretching out to 400 miles (644 kilometers) north toward the New Hebrides.[30] Despite no confirmed sightings of Japanese submarines or warships, the experience proved beneficial in that it gave the New Zealanders and Americans an opportunity to live, train, and work together. USAAF units in particular demonstrated a spirit of cooperation—for example, by providing access to spares in that parts of the Cyclone engine that powered the B-17 and parts of the Twin Wasp engine that powered the Hudson were interchangeable. In return, the New Zealanders took over some ground duties, such as refueling aircraft and operating crash tenders. USAAF and RNZAF aircrews took the opportunity to practice antishipping strikes together.[31]

The Americans appeared reluctant to utilize RNZAF squadrons in the developing Solomons campaign. However, Goddard predicted that this situation would change. He reasoned that competing demands from the Europe–Middle East and Asia-Pacific theaters would mean operations in the SOPAC were shaped by "economy of force and resources."[32] Indeed, the JCS informed Ghormley this was the case.[33] Goddard suggested that because the RNZAF personnel were well trained and shared with their American comrades a "common language and, in general, common doctrine," interoperability would be achievable provided RNZAF squadrons could be suitably equipped.[34] Within a short time after the counteroffensive was launched in August 1942, the Americans showed more interest in working with the RNZAF. The air war over Guadalcanal stretched USN, USMC, and USAAF squadrons. Faced with an ever-increasing demand for strategic and tactical air power, McCain considered how he might utilize the RNZAF. In late September, he handed over the role of COMAIRSOPAC to Rear Admiral Aubrey W. Fitch, USN, who had recently commanded an aircraft carrier task force in the Battle of the Coral Sea. Returning stateside, McCain met with Nimitz, who asked how they might employ "Goddard's outfit." McCain informed Nimitz that the New Zealanders, if suitably reequipped, could man a heavy-bomber squadron and some medium-bomber and fighter squadrons.[35]

Unaware of this dialogue, Goddard drafted a discussion paper that he sent out to the senior echelons of the RNZAF, the New Zealand government, and SOPAC in October 1942, complaining that there was "no spontaneous planning in higher United States Staffs, for the development of the Royal New Zealand Air Force to meet future requirements in the South Pacific."[36] Goddard advocated expanding the RNZAF to thirty squadrons to form a "balanced, offensive, mobile air fleet," operating a mix of heavy bombers, torpedo bombers, dive bombers, fighters, transports, and flying boats. In essence, his contention was that the RNZAF should be capable of more than hunting and stalking the enemy at sea, noting, "It is in the offensive role that the fighting qualities of New Zealanders have been best exemplified. It is considered that these qualities should be utilised."[37] Group Captain A. T. Nevill, Deputy CAS (DCAS), delivered Goddard's paper to SOPAC headquarters in the New Hebrides. Ghormley was under intense pressure at the time, having suffered a series of setbacks in the Solomons.[38] On 13 October, he sent a short note to Goddard acknowledging that he had yet to read the paper but promising to talk with his air and land commanders about it at some point.[39] Goddard never received a further response because within two days Nimitz decided that the SOPAC required a "more aggressive commander." He replaced Ghormley with Vice Admiral William F. Halsey, USN.[40]

INCREASING THE RNZAF CONTRIBUTION

Although relations with Ghormley had been cordial enough, New Zealand's government had become dissatisfied with the lack of communication from the Americans and their seeming reluctance to utilize New Zealand forces. After Halsey settled into his command, Minister of Armed Forces and War Coordination Gordon Coates led a delegation to his headquarters. Halsey had revealed a capacity for joint command, pulling his naval, marine, army, and air commanders into a cohesive team and inspiring them to achieve the desired results.[41] He also showed a capacity for multinational command, both welcoming the New Zealanders and readily agreeing to their request to have an NZCS representative appointed to his staff.[42] This response was characteristic of Halsey. Whereas MacArthur's nature and command style in the SWPA proved divisive, Halsey possessed the personality and professionalism required to produce a unified command. As one of his biographers, James M. Merrill, explains, "Halsey insisted on exercising unity of command over all forces in the South Pacific and enforced this concept right on down."[43] Halsey was already planning to utilize New Zealand forces, albeit in a modest fashion. His COMAIRSOPAC, Fitch, had authorized the attaching of some Fiji-based flying-boat crews to an American squadron and organized for a New Zealand-based fighter squadron and another general-reconnaissance squadron to be deployed to islands in the rear area. On 6 November 1942, Fitch wrote to Goddard that he could think of "no immediate further avenue of usefulness of New Zealand Air Force units in this sector" but reassured him that "further needs for your personnel may and probably will become apparent."[44]

The deploying of a fighter squadron to the SOPAC came about after the JCS proposed that the RNZAF could take over the aircraft and equipment of up to six USAAF squadrons to enable a reduction in the number of aircraft sent to New Zealand at a time when shipping was in short supply. The initiative was certainly worth trying. By mid-1942, the RNZAF had an oversupply of fighter pilots, including some with operational experience in Malaya, but a dire shortage of fighters. It had enough Kittyhawks to form an operational training unit and three of five planned fighter squadrons.[45] The Americans meanwhile had a pressing need for fighters at Guadalcanal. Major General Millard F. Harmon, commanding general, US Army Forces in the SOPAC, proposed to relieve the USAAF's 68th Fighter Squadron from garrison duties on the out-the-way island of Tonga. By having a RNZAF squadron take over its P-40s and equipment, the 68th could be shipped to New Caledonia, reequipped, and sent to Guadalcanal, while the RNZAF squadron could leave its aircraft in New Zealand for a new squadron. Goddard agreed but disapproved of the idea that cast-off P-40s would be equivalent to new P-40s.[46] Number 15 Squadron, RNZAF, sailed for Tonga in October 1942 and took possession of the Americans' aircraft, spares, and munitions. As Goddard evidently had feared, the airframes, engines, and equipment were in poor condition.[47] Moreover, "defending" an island that had never experienced an air raid proved a thankless task. Group Captain Sidney Wallingford, the RNZAF representative with the COMAIRSOPAC headquarters, observed that "they were inclined to be a recalcitrant lot" on Tonga.[48]

Harmon also requested another Hudson squadron for New Caledonia. This request would stretch the RNZAF, which was short of ground staff able to serve in the tropics. Recruitment policy directed the majority of A-grade physically fit men into the army, leaving the RNZAF with many B-grade men restricted to serving in New Zealand.[49] The RNZAF scoured air force stations across the country to identify 300 fit men and rushed to outfit them with tropical uniforms and kit. By the time Number 3 (General Reconnaissance) Squadron assembled and shipping of its ground staff, spare aircrews, and equipment was arranged, Fitch decided the squadron should proceed to the New Hebrides to support operations at Guadalcanal.[50] Having recognized that the RNZAF was able to provide a niche maritime patrol capability, the Americans entrusted the squadron to patrol over shipping lanes leading to and from Guadalcanal. The squadron started operating from Espiritu Santo in October 1942, and the next month sent a detachment of six Hudsons to Guadalcanal, where an armed maritime reconnaissance capability was badly needed. The "Cactus Air Force" had made do with USMC dive-bomber crews pressed into this role. The New Zealanders took over reconnaissance of the seas and island coastlines between Guadalcanal and the Japanese base at Rabaul, New Britain, searching for warships and barges as well as on occasion for downed Allied aircrews. The demand for sorties was such that the remainder of the squadron was called forward in December.[51] Wallingford observed that the Hudson crews' "steady and efficient plodding . . . really made the reputation here for the RNZAF."[52]

Operation Cartwheel

In the wake of hard-won victories at Guadalcanal and at Buna in New Guinea in January–February 1943, the JCS instructed the SOPAC and SWPA commands to undertake Operation Cartwheel. Under MacArthur's overall command, forces in the Solomons and New Guinea were to push toward, isolate, and converge on Rabaul. For his part, Halsey planned a series of amphibious landings to inch closer to Rabaul and seize ground for new airfields—with each operation extending the operating radius of his air forces. The air war over Guadalcanal had depleted the naval, marine, and army squadrons, with aircrews and ground staff showing clear signs of fatigue.[53] With the squadrons' efficiency declining, Halsey and Fitch began advocating for use of the RNZAF. The JCS approved building up the RNZAF to fifteen squadrons, while the NZCS agreed to release most home-based squadrons for service in "the islands."[54] Anticipating a sharp increase in the number of units deployed, Goddard had Wallingford extracted from the COMAIRSOPAC headquarters to form and command a new administrative formation, Number 1 (Islands) Group, RNZAF, based on Espiritu Santo. Promoted to air commodore, Wallingford would liaise with the COMAIRSOPAC and oversee the administration and logistic support of RNZAF squadrons and ancillary units so that they could be slotted into American-commanded task forces.[55]

In April 1943, Fitch established a new command, Air, Solomons (AIRSOLS), to support the advance in the Solomons and contribute to the bombing of Rabaul. Rear Admiral Charles P. Mason, USN, a naval aviator since the First World War, was appointed to the command of AIRSOLS. Mason had responsibility for an "astonishing mixture" of aircraft belonging to the USN, USMC, USAAF, and RNZAF. The task of "welding this conglomerate air force into a smoothly functioning organization was not an easy one nor was it achieved at once," but Mason and his successors were ultimately successful.[56] For the New Zealanders, an important factor in the relationship was the unity of command insisted upon by Halsey and exercised in AIRSOLS. A senior British observer who visited the SOPAC in early 1944 noted that the American–New Zealand relationship was "much closer and more intimate" than the at times strained relationship between the Australians and Americans in the SWPA.[57]

The nature and extent of New Zealand air operations evolved following the formation of AIRSOLS. A significant development was the deployment of fighters, starting with the transfer of Number 15 Squadron from Tonga to Guadalcanal at the end of April. The RNZAF also offered up a radar unit at this time. During 1942, the RNZAF received radar equipment from Britain to establish a home-defense early-air-warning network. The equipment was jealously guarded while there was still a fear of air raids. In February 1943, the NZCS admonished the director of the radar program, Dr. Ernest Marsden, for offering Halsey the use of a radar set and personnel. Halsey naturally welcomed the offer. The NZCS was more or less obliged to start releasing radar units for service in the South Pacific as an overdue gesture of reciprocity.[58] Number 52 Radar Unit worked with USN, USMC, and USAAF night fighters engaged in countering "Washing Machine Charlie"—the collective noun for Japanese bombers that harassed Guadalcanal at night—and proved instrumental in detecting daylight raids, which continued until June 1943, when Japanese air raids began petering out.[59]

The high operational tempo expected of fighter squadrons and Guadalcanal's enervating climate conspired to reduce squadron efficiency within a matter of weeks. Number 1 (Islands) Group determined that fighter squadrons should be rotated approximately every six weeks, in contrast to the system for Hudson aircrews, who were liable to remain with their squadron until showing signs of fatigue, meaning in effect that "the chaps who could take it, remained the longest."[60] Ground staff did not suffer the same strains and were found to be efficient for longer. Number 1 (Islands) Group responded by pulling technical ground staff out of squadrons to form mobile servicing units, which could be sent to forward airfields with one squadron and remain there to service the replacement squadron.[61] In addition, from February 1943 there was greater integration of ground staff into the American organization. Number 1 (Islands) Group lent several dozen air mechanics to USN workshops in the New Hebrides to assist with servicing Grumman F4F Wildcat fighters and Douglas SBD Dauntless dive bombers.[62]

RNZAF Kittyhawk squadrons demonstrated their worth with a respectable tally of downed enemy aircraft. The fighter wing was to attain ninety-nine confirmed aerial victories in 1943. The New Zealanders' efforts were made even more important by the fact that American fighter squadrons were close to exhaustion in the middle of the year.[63] In July 1943, Fitch proposed to have RNZAF pilots posted as reinforcements to US naval and marine fighter squadrons, which indicated his level of confidence in New Zealand fighter training.[64] Wallingford railed against the idea, however, noting that

RNZAF pilots were not familiar with the Wildcat equipping most of the naval and marine fighter squadrons.[65] Moreover, these squadrons were in the process of transitioning to Grumman F6F Hellcats and Chance-Vought F4U Corsairs, respectively. When knocked back, Fitch offered the RNZAF enough surplus Wildcats to form two fighter squadrons, in addition to the five Kittyhawk squadrons.[66] Air Headquarters (the RNZAF's administrative headquarters in Wellington, New Zealand) briefly entertained the idea but informed Fitch that a shortage of trained ground staff meant the initiative was not viable. The shortage of ground staff was real, but senior officers also worried that the RNZAF could be stuck with obsolescent and worn-out Wildcats.[67]

The offer of naval fighters marked a significant development in the equipping of the RNZAF. In June 1943, the USN assumed responsibility for the logistic support of squadrons operating USN-type aircraft, while the USAAF supported those with US Army–type aircraft. During 1941–1942, the majority of aircraft delivered to the RNZAF were US Army types, in particular the A-29 Hudson and the P-40 Warhawk/Kittyhawk. Six months after P-40E deliveries stopped in mid-1942, newer models of the P-40 arrived to equip fighter squadrons into 1944. The Hudson, however, reached the end of its operational service life by the middle of 1943. The RNZAF accepted the Lockheed B-34 Lexington as a replacement. The Lexington was reported to have "a good reputation as an offensive bombing type," enabling the general-reconnaissance squadrons to be redesignated as bomber-reconnaissance squadrons.[68] The first Lexingtons reached New Zealand in June 1943, but after only one squadron was equipped, the supply was switched to the naval variant, the PV-1 Ventura. USAAF squadrons were starting to be transferred from the SOPAC to the SWPA, leaving the USN as the dominant air force in the SOPAC. It made sense for the USN to enhance interoperability by supplying the New Zealanders with its own aircraft types and taking on their logistic support. The switch from the B-34 to the PV-1 does not appear to have concerned the senior echelons of the RNZAF. Their main discussion was around the fact that both variants could be utilized in an offensive role. Wallingford noted that experience with Hudsons had shown that aircrews welcomed the ability to have "a crack at the Japs by way of relief from some of the rather monotonous patrols."[69] Air Vice Marshal Leonard Isitt, who replaced Goddard in 1943 as the first New Zealand–born CAS, agreed that although maritime reconnaissance remained vital, the fact that the new aircraft could engage in bombing as well could only be good news. He noted that "the odd strike or bomb raid has an amazing effect on the outlook of the crews."[70]

The push toward naval aircraft types became even more apparent with allocations of dive bombers and torpedo bombers in the second half of 1943. At the end of July 1943, Number 25 Squadron became the RNZAF's first and only dive-bomber squadron when it took over "war weary" SBD Dauntlesses from the USMC. The aircraft were in such poor condition that ground staff had to "cannibalize" several to make a handful flyable.[71] Operational training continued until February 1944, when the squadron deployed to Guadalcanal with new Dauntlesses. It was joined by the first of two squadrons to be equipped with Grumman TBF Avenger torpedo bombers. A lack of shipping targets in the SOPAC meant the need for torpedo bombers had passed, so the New Zealand aircrews trained alongside Americans to master dive-bombing and glide-bombing formations and tactics. Combined training enabled the New Zealanders to work closely with USMC and USN Dauntless and Avenger squadrons and to engage in close air support of American troops on Bougainville. The New Zealanders regularly joined large-formation American raids against Rabaul and airfields elsewhere in New Britain and New Ireland.[72]

Although Dauntless and Avenger operations generally proved effective, those squadrons that operated these two types of aircraft would have a short lifespan. Even before the squadrons had entered the operational area, the RNZAF was proposing to streamline the squadrons' equipment. On 17 September 1943, Isitt wrote to Fitch explaining that although the JCS approved supply for nineteen New Zealand squadrons, under existing arrangements the squadrons would be operating in six different roles (bomber reconnaissance, fighter, dive bomber, torpedo bomber, flying boat, and transport) as well as with eight or nine different types of aircraft and five main types of engine. This variability, he suggested, was unsustainable: "The multiplicity of types contained in an Air Force of comparatively small size tends to increase beyond economic limits the overheads in the shape of repair and maintenance facilities, stores, etc., and necessitate[s] the transportation from the United States of more squadron basic equipment than would be required for a force of similar size of less variety of types."[73]

Noting that the New Zealand government desired the RNZAF to "participate to the maximum extent in offensive operations," Isitt suggested discarding dive bombers and torpedo bombers, whose utility he doubted, and reorganizing the air force with a narrower range of roles: six medium bomber or bomber-reconnaissance squadrons, ten fighter squadrons, two flying-boat squadrons, and one transport squadron.[74] Such a change required the approval of the JCS; however, this approval was slow in coming. At the end of 1943, Isitt's deputy, Nevill, expressed the frustration that he could offer only "the wildest guess" as to the nature and extent of RNZAF operations in the coming year because "we have no idea what our aircraft allotments for 1944 are going to be, or when they are likely to be delivered."[75] In January 1944, the JCS approved Isitt's plan, with only the slight variation of allowing for two transport squadrons. Fighter squadrons were to be equipped with F4U Corsairs (replacing Kittyhawks), bomber-reconnaissance squadrons with PV-1 Venturas, flying-boat squadrons with PBY Catalinas, and transport squadrons with Douglas C-47 Dakotas and Lockheed C-60 Lodestars.[76]

Other pressures also came to influence New Zealand and American planning for utilization of the RNZAF. New Zealand's manpower reserve was stretched, and it became harder to sustain forces in both the European and Pacific theaters. The government opted to disband the 3rd New Zealand Division, which had served in the SOPAC for a year, to channel the majority of army recruits into the 2nd Division in Italy. It determined, however, that the RNZAF should continue operating in the Pacific. Unfortunately, with the Solomons campaign then winding down, there were signs that American interest in the RNZAF was waning. Air Commodore M. W. Buckley, who had taken over the command of Number 1 (Islands) Group, complained to Fitch in February 1944 that SOPAC plans for future operations made "no specific mention of RNZAF units."[77] Fitch responded that the SOPAC had planned for the "use of all types [of] RNZAF units," but with the Japanese forces in the Solomons contained, it was impossible to predict where any Allied squadrons in the theater would be employed in the future. He anticipated the continuance of air strikes to harass isolated Japanese garrisons but could not commit to utilizing the New Zealanders.[78] Buckley recommended to Air Headquarters in Wellington that no further RNZAF units be deployed "until plans for their employment have been formulated," but senior officers wanted to avoid the "embarrassment" of having trained units staying on the home front, essentially "unemployed," because this would call into question the RNZAF's manpower allocations. Nevill informed Buckley that the senior officers would "rather see them go forward in anticipation of full employment."[79]

In the forward area, there was unease at the changing nature of air operations in the SOPAC. This was most evident in the case of fighter squadrons.

Although New Zealand squadrons had achieved a fine record as aerial fighters, once air superiority was achieved, there was little use for the fighter squadrons except for in a ground-attack role.[80] Close air support became vital to Allied operations in the last years of the war, but the emphasis on these sorties was liable to be a cause of tension because they were not as glamorous as fighter interdiction.[81] Buckley believed it was rather pointless to deploy fighter squadrons "for sporadic misuse as bombers."[82] Nevill rejected this take on the situation, however, noting that squadrons were "doing good work as fighter bombers." He maintained that the Kittyhawk had proved "particularly valuable in this role elsewhere, and I hardly regard it as a misuse of this type if the situation permits such employment."[83]

TO THE SOUTHWEST PACIFIC AREA

The reduction of US forces in the SOPAC gave rise to a concern that the Americans would not continue utilizing and supporting RNZAF squadrons. In April 1944, a JCS options paper signaled that the RNZAF could be relegated to garrison duties on bypassed islands. This was not the offensive role desired by the New Zealanders, however. Isitt labeled the suggestion "most unsatisfactory."[84] Conscious of the New Zealand government's war aims and expectations regarding the use of the RNZAF, he wrote to Prime Minister Fraser to reassure him that the air force's "object is to press for employment in the combat zone and it is still more important that plans should be developed for the ultimate employment of the RNZAF in active operations against Japan in order that adequate representation is secured at the termination of hostilities."[85] In response to the criticism from New Zealand, the JCS authorized the transfer of several RNZAF squadrons to the SWPA not only to enable the New Zealanders to play a garrison role in the SOPAC but also to give them an opportunity to continue offensive operations.

In August 1944, SOPAC, SWPA, and RNZAF representatives met on New Caledonia to discuss the transfer of four RNZAF fighter squadrons and two bomber-reconnaissance squadrons to the SWPA. An RNZAF historical summary prepared in this period indicated that both MacArthur and his air commander, Lieutenant General George C. Kenney, welcomed news of the transfer. The report's author hoped that "the good relations which prevailed between the United States Navy and the RNZAF will continue with our new masters in the South-west Pacific."[86] This was not likely, however. Kenney had a very different command style than the COMAIRSOPAC. Kenney preferred cooperation between the services rather than a unified command. For Isitt, the most important question was where the six squadrons sent to the SWPA could be employed. He hoped they could support the American campaign in the Philippines, but when he visited MacArthur's headquarters in Brisbane, Australia, to further discuss the transfer, Kenney broke the news that the New Zealanders would support Australian forces elsewhere. MacArthur had instructed the commander of the Australian Military Forces, General Thomas Blamey, to deploy infantry divisions to northern New Guinea, New Britain, and Bougainville. MacArthur in effect sidelined the Australians by relegating them to "mopping up" operations against isolated garrisons posing no real threat. Blamey planned a series of limited offensives with the objective of keeping casualties to a minimum.[87]

Although disappointed to be relegated to a backwater, senior RNZAF officers held out hope that the squadrons might later be transferred westward to support MacArthur's main offensive. It turned out that the Americans did not anticipate such a transfer because of the aircraft that equipped the New Zealand squadrons. Wing Commander G. H. Pirie, the RNZAF's director of operations, spoke with Kenney, who acknowledged that Bougainville "might not present the most active form of employment" but then explained that deploying the RNZAF squadrons there would "simplify" their logistic support because neither the 5th nor the 13th Air Forces under his command were in a position to support naval aircraft. Kenney's chief of staff, Colonel R. E. Beebe, emphasized that the RNZAF should lobby the JCS to be reequipped with US Army aircraft types if it were to have any chance of having squadrons in the Philippines. He recommended North American P-51 Mustangs to replace the Corsairs and either the Douglas A-20 Havoc or Douglas A-26 Invader or North American B-25 Mitchell to replace the Venturas. Pirie noted that complete reequipping would be needed; otherwise, the small air force would be presented with "an almost insuperable problem" of having to operate, logistically support, and train aircrews and ground staff to operate a broad mix of aircraft types on operations and on garrison duties.[88]

Determined that New Zealand would contribute to the Pacific War until the defeat of Japan, the government agreed to the deployment of a New Zealand Air Task Force to Bougainville. During October–December 1944, the Australians took over land operations on the island from the Americans. By this time, the RNZAF had thirteen fighter-bomber squadrons with Corsairs, six bomber-reconnaissance squadrons with Venturas, two patrol squadrons with Catalinas, and two transport squadrons with a mix of Dakotas and Lodestars. (This contribution to the war effort was in addition to but not as well known as the seven New Zealand squadrons with the RAF in Europe.) Two Corsair squadrons reached Bougainville in late 1944, and the task force was then doubled in strength, with the Corsairs used for air strikes against Japanese bases and lines of communication as well as for close air support of Australian troops. Regular rotations of squadrons meant tours generally lasted around two months. Bomber-reconnaissance squadrons meanwhile mainly flew air strikes over New Britain; however, by mid-1945 four Ventura-equipped squadrons were disbanded as the need for bombing sorties was reduced. At the war's end, the RNZAF had nine Corsair squadrons in forward areas, including Bougainville, Los Negros, and Jacquinot Bay in New Britain.[89] Although equipping the RNZAF with USN-type aircraft had imposed limitations in 1944–1945, the RNZAF nevertheless met its aim of contributing to the war against Japan until its end. The CAS, Air Vice Marshal Isitt, was appropriately selected to represent New Zealand at the Japanese surrender in Tokyo Bay on 2 September 1945.

Conclusion

Interoperability offers advantages for coalition partners but often comes at a price.[90] For the RNZAF, the price was realized only toward the end of the Pacific War. The American equipping of New Zealand squadrons was of tremendous assistance, particularly during 1942–1943, when the RNZAF sought to make a material contribution to the war against Japan. The common language and a compatible doctrine no doubt eased the challenge of cooperation. Reequipping the RNZAF ensured that a small and ill-equipped air force could emerge as a regional air power, albeit as the junior partner in an American-dominated force. This process had obvious benefits for both parties in 1943, with the New Zealanders able to sustain maritime patrol operations and boost fighter operations in the Solomons when the Americans needed such support. However, because the dominant service in the SOPAC influenced the equipping process, the RNZAF came to be equipped almost entirely with US Navy–type aircraft. This had implications in the last year of the war, when MacArthur's command pointed to the operation of naval aircraft as a reason, or excuse, for a political decision to sideline the New Zealanders alongside the Australians in the Bougainville campaign.

Notes

1. At the start of the Second World War, the British Empire's air forces were the Royal Air Force (established in 1918), the Royal Canadian Air Force (1920), the South African Air Force (1920), the Royal Australian Air Force (1921), and the Royal New Zealand Air Force (1937). The Irish Air Corps (1924) was a branch of Eire's Army.

2. Ashley Jackson, *The British Empire and the Second World War* (London: Hambledon Continuum, 2006), 487.

3. J. M. S. Ross, *Royal New Zealand Air Force,* vol. in *Official History of New Zealand in the Second World War 1939–45* (Wellington: War History Branch, New Zealand Department of Internal Affairs, 1955).

4. Ian McGibbon, "'Something of Them Is Here Recorded': Official History in New Zealand," in *The Last Word? Essays on Official History in the United States and British Commonwealth,* ed. Jeffrey Grey (Westport, CT: Praeger, 2003), 53.

5. See in particular Ross, *Royal New Zealand Air Force,* 319–20.

6. Brian J. Hewson, "Goliath's Apprentice: The Royal New Zealand Air Force and the United States in the Pacific War 1941–1945," PhD diss., University of Canterbury, 1999, 303.

7. Wesley Frank Craven and James Lea Cate, *The Pacific: Guadalcanal to Saipan, August 1942 to July 1944,* vol. 4 of *The Army Air Forces in World War II,* ed. Wesley Frank Craven and James Lea Cate (Chicago: University of Chicago Press, 1950), 89.

8. See Roger H. Palin, *Multinational Military Forces: Problems and Prospects* (Abingdon, UK: Routledge, 1995), 39–40.

9. Simon Moody, "Reflecting on the Bettington Report of 1919—a Centennial Legacy," *Journal of the Royal New Zealand Air Force* 5, no. 1 (2019): 55–64.

10. Paul Stockley, "The RNZAF as an Independent Air Force: Then and Now," *Journal of the Royal New Zealand Air Force* 4, no. 1 (2018): 62.

11. Ross, *Royal New Zealand Air Force,* 25–26.

12. Fred Jones, Minister for Defence, speech to New Zealand House of Representatives, in *New Zealand Parliamentary Debates* (Hansard), 19 July 1939 (Wellington: New Zealand Government Printer, 1940), 563.

13. S. D. Waters, *The Royal New Zealand Navy,* vol. in *Official History of New Zealand in the Second World War 1939–45* (Wellington: Historical Publications Branch, New Zealand Department of Internal Affairs, 1956), 138–40.

14. O. A. Gillespie, *The Pacific,* vol. in *Official History of New Zealand in the Second World War 1939–45* (Wellington: War History Branch, New Zealand Department of Internal Affairs, 1952), 19–20; Ross, *Royal New Zealand Air Force,* 71–72.

15. Governor-General of New Zealand to Secretary for Dominion Affairs, London, conveying message from Prime Minister Peter Fraser to British Prime Minister Winston Churchill, 4 December 1940, doc. 191, in *Documents Relating to New Zealand's Participation in the Second World War 1939–45,* vol. 3 (Wellington: War History Branch, New Zealand Department of Internal Affairs, 1963), 214–15.

16. Australian Department of Defence Coordination to New Zealand Prime Minister's Department, telegram, 17 January 1941, A2676, 700, National Archives of Australia (NAA), Canberra.

17. See Louis Morton, *The Fall of the Philippines*, vol. in *United States Army in World War II: The War in the Pacific* (Washington, DC: Office of the Chief of Military History, US Department of the Army, 1953), 37–38; Wesley Frank Craven and James Lea Cate, *Services around the World*, vol. 7 of *The Army Air Forces in World War II*, ed. Wesley Frank Craven and James Lea Cate (Chicago: University of Chicago Press, 1958), 173–74.

18. Wesley Frank Craven and James Lea Cate, *Plans and Early Operations, January 1939 to August 1942*, vol. 1 of *The Army Air Forces in World War II*, ed. Wesley Frank Craven and James Lea Cate (Chicago: University of Chicago Press, 1948), 180–81.

19. One of the Singapores was on its delivery flight from Singapore (Ross, *Royal New Zealand Air Force*, 109).

20. Ibid., 109–11.

21. Craven and Cate, *Plans and Early Operations*, 430–31.

22. Robert Lowry, *Fortress Fiji: Holding the Line in the Pacific War, 1939–1945* (Sutton, UK: R. W. Lowry, 2006), 33–34.

23. Ross, *Royal New Zealand Air Force*, 125–26.

24. John Miller Jr., *Guadalcanal: The First Offensive*, vol. in *United States Army in World War II: The War in the Pacific* (Washington, DC: Historical Division, US Department of the Army, 1949), 2–3.

25. Ross, *Royal New Zealand Air Force*, 126; Hewson, "Goliath's Apprentice," 62.

26. New Zealand Prime Minister to New Zealand Minister in Washington, DC, draft cablegram (marked "Not Sent"), 12 July 1942, quoted in Hewson, "Goliath's Apprentice," 84.

27. Ross, *Royal New Zealand Air Force*, 126–28.

28. Kramer J. Rohfleisch, *The AAF in the South Pacific to October 1942*, US Air Force Historical Study no. 101 (Washington, DC: Assistant Chief of Air Staff, Intelligence, Historical Division, USAAF, December 1944), 28–40.

29. Ibid., 29–30; Alton Keith Gilbert, *A Leader Born: The Life of Admiral John Sidney McCain, Pacific Carrier Commander* (Drexel Hill, PA: Casemate, 2006), 73–79.

30. Miller, *Guadalcanal*, 32–33.

31. Ross, *Royal New Zealand Air Force*, 131–33.

32. Air Commodore Victor Goddard, CAS, RNZAF, "Proposals for Development for Participation in South Pacific War Operations," 9 October 1942, AIR1, 949/130/19/1 Part 1, Archives New Zealand (ANZ), Wellington.

33. Samuel Eliot Morison, *Breaking the Bismarcks Barrier*, vol. 6 of *History of United States Naval Operations in World War II* (Boston: Little, Brown, 1950), 93.

34. Goddard, "Proposals."

35. Hewson, "Goliath's Apprentice," 99.

36. Goddard, "Proposals."

37. Ibid.

38. Samuel Eliot Morison, *The Struggle for Guadalcanal, August 1942–February 1943*, vol. 5 of *History of United States Naval Operations in World War II* (Boston: Little, Brown, 1949), 182.

39. Admiral Robert L. Ghormley to Air Commodore Victor Goddard, 13 October 1942, AIR1, 949/130/19/1 Part 1, ANZ, Wellington.

40. Morison, *The Struggle for Guadalcanal*, 183.

41. Thomas Alexander Hughes, *Admiral Bill Halsey: A Naval Life* (Cambridge, MA: Harvard University Press, 2016), 187–97.

42. Hewson, "Goliath's Apprentice," 108–10.

43. James M. Merrill, *A Sailor's Admiral: A Biography of William F. Halsey* (New York: Crowell, 1976), 84. For the best assessment of MacArthur's command, see Peter J. Dean, *MacArthur's Coalition: US and Australian Military Operations in the Southwest Pacific Area, 1942–1945* (Lawrence: University Press of Kansas, 2018).

44. Rear Admiral Aubrey W. Fitch, COMAIRSOPAC, to Goddard, 6 November 1942, AIR1, 949/130/19/1 Part 1, ANZ, Wellington.

45. Air Headquarters, Wellington, to New Zealand Air Attaché, Washington, DC, cable, 26 October 1942, AIR1, 949/130/19/1 Part 1, ANZ, Wellington.

46. Hewson, "Goliath's Apprentice," 102.

47. Ross, *Royal New Zealand Air Force,* 139–41.

48. Air Commodore Sidney Wallingford, Air Officer Commanding, No. 1 (Islands) Group, RNZAF, to Air Commodore L. M. Isitt, Deputy Chief of Air Staff, RNZAF, 4 June 1943, AIR100, 1/4, ANZ, Wellington.

49. Hewson, "Goliath's Apprentice," 88.

50. Ross, *Royal New Zealand Air Force,* 137–38.

51. Ibid., 147–48.

52. Wallingford to Isitt, 4 June 1943.

53. Craven and Cate, *The Pacific,* 275.

54. Air Department to New Zealand Chiefs of Staff, "Deployment of the RNZAF and Allocation of Squadrons for Local Defence," memorandum, 23 February 1943, appendix B, "Draft Letter to the Commander, South Pacific Area," AIR1, 949/130/19/1 Part 1, ANZ, Wellington.

55. Ross, *Royal New Zealand Air Force,* 160.

56. Craven and Cate, *The Pacific,* 89.

57. Major-General R. H. Dewing, "Report on Visit to South Pacific Area," 1942, 706.04 (digitized, IRIS no. A7135), US Air Force Historical Research Agency.

58. Ross Galbreath, "New Zealand Scientists in Action: The Radio Development Laboratory and the Pacific War," in *Science and the Pacific War: Science and Survival in the Pacific, 1939–1945,* ed. Roy M. Macleod (Dordrecht, Netherlands: Kluwer Academic, 2000), 218–19.

59. William N. Ness, *Pacific Sweep: The 5th and 13th Fighter Commands in World War II* (New York: Doubleday, 1974), 68–69; Ross, *Royal New Zealand Air Force,* 230–31.

60. Director of Operations, RNZAF, to DCAS, "Bomber Reconnaissance Squadrons— Policy and Rotation," 14 July 1943, AIR1, 949/130/19/1 Part 1, ANZ, Wellington.

61. Ross, *Royal New Zealand Air Force,* 166–67.

62. Ibid., 64–69.

63. Henry I. Shaw Jr. and Douglas T. Kane, *Isolation of Rabaul,* vol. 2 of *History of US Marine Corps Operations in World War II* (Washington, DC: Historical Branch, US Marine Corps, 1963), 466–67.

64. Colonel C. W. Salmon, New Zealand Chiefs of Staff Representative, SOPAC, to Army Headquarters, Wellington, for forwarding to Goddard, telegram, 5 July 1943, AIR1, 949/130/19/1 Part 1, ANZ, Wellington.

65. Wallingford to Goddard, telegram, 5 July 1943, AIR1, 949/130/19/1 Part 1, ANZ, Wellington.

66. COMAIRSOPAC to Air Department, Wellington, telegram, 5 August 1943, AIR1, 949/130/19/1 Part 1, ANZ, Wellington.

67. Air Department to COMAIRSOPAC, 5 August 1943, AIR1, 949/130/19/1 Part 1, ANZ, Wellington.

68. Director of Operations, RNZAF, to DCAS, "Bomber Reconnaissance Squadrons—Policy and Rotation," 14 July 1943.

69. Wallingford to Isitt, CAS, 30 July 1943, AIR100, 1/4, ANZ, Wellington.

70. Isitt to Wallingford, 3 August 1943, AIR100, 1/4, ANZ, Wellington.

71. Cliff F. L. Jenks and Phil H. Listemann, *The Dauntless in RNZAF Service* (n.p.: n.p., 2015), 5–6.

72. Ross, *Royal New Zealand Air Force*, 247–51.

73. Isitt, CAS, to COMAIRSOPAC, "RNZAF: Future Organization and Employment," 17 September 1944, AIR1, 949/130/19/1 Part 1, ANZ, Wellington.

74. Ibid.

75. Group Captain A. T. Nevill to Director of Operations, RNZAF, "Deployment of RNZAF Units during 1944," 13 December 1943, AIR1, 949/130/19/1 Part 2, ANZ, Wellington.

76. Isitt to COMAIRSOPAC, "Plan for Future Deployment and Employment of the RNZAF in the South Pacific," 29 February 1944, AIR1, 949/130/19/1 Part 2, ANZ, Wellington.

77. Air Commodore M. W. Buckley, No. 1 (Islands) Group, RNZAF, to COMAIR-SOPAC, "Plan for Future Deployment and Employment of the RNZAF in the South Pacific Area," 29 February 1944, AIR1, 949/130/19/1 Part 1, ANZ, Wellington.

78. COMAIRSOPAC to Buckley, "Plans for Future Deployment and Employment of Royal New Zealand Air Force in the South Pacific Area," 13 March 1944, AIR1, 949/130/19/1 Part 2, ANZ, Wellington.

79. Buckley to Air Department, "Plans for the Future Deployment and Employment of the RNZAF in the South Pacific," 16 March 1944, AIR1, 949/130/19/1 Part 2, ANZ; Nevill to Buckley, "RNZAF Planning in South Pacific Area," 22 March 1944, AIR1, 949/130/19/1 Part 2, ANZ, Wellington.

80. See Ross, *Royal New Zealand Air Force*, 175–97.

81. Joe Gray Taylor, "American Experience in the Southwest Pacific," in *Case Studies in the Development of Close Air Support,* ed. Benjamin Franklin Cooling (Washington, DC: Office of Air Force History, US Air Force, 1990), 315.

82. Buckley to Air Department, "Plans for the Future Deployment and Employment of the RNZAF in the South Pacific," 16 March 1944.

83. Nevill to Buckley, "RNZAF Planning in South Pacific Area," 22 March 1944.

84. Isitt, CAS, RNZAF, to RNZAF Liaison Officer, Washington, DC, 4 April 1944, AIR1, 949/130/19/1 Part 2, ANZ, Wellington.

85. Isitt to Prime Minister Fraser, "Operational Employment of RNZAF in the Pacific Theatre," 5 April 1944, AIR1, 949/130/19/1 Part 2, ANZ, Wellington.

86. Ibid.

87. Karl James, *The Hard Slog: Australians in the Bougainville Campaign, 1944–45* (Melbourne: Cambridge University Press, 2014), 16–20.

88. Wing Commander G. H. Pirie, RAF (seconded to RNZAF), Director of Operations, RNZAF, "Deployment of RNZAF Squadrons in South-West Pacific Areas," 8 August 1944, AIR1, 949/130/19/1 Part 1, ANZ, Wellington.

89. Ross, *Royal New Zealand Air Force*, 291–311.

90. Myron Hura, Gary W. McLeod, Eric V. Larson, James Schneider, Daniel Gonzales, Daniel M. Norton, Jody Jacobs, Kevin M. O'Connell, William Little, Richard Mesic, and Lewis Jamison, *Interoperability: A Continuing Challenge in Coalition Air Operations* (Santa Monica, CA: RAND, 2000), 7.

From Ally to Auxiliary

The Royal Hungarian Air Force and the Luftwaffe in the Second World War

Stephen L. Renner

Multinational military cooperation is seldom a story of perfect peers sharing equally their contributions and bearing equally the risks. The chapters of this book describe many cases in which one member of the alliance is or becomes dominant in the course of a conflict. But in none of the other examples does one nation enter a conflict as a sovereign, if reluctant, partner only to find itself a vassal of its former ally before the war ends. That was Hungary's experience in the Second World War. Hungarian foreign policy after the First World War was dominated by the desire to reverse the losses suffered at the Paris peace table (two-thirds of its prewar territory, nearly half of its population), and Budapest found that any significant revision in its favor would come only from Berlin.

Hungarian airmen between the wars enjoyed a robust debate about the proper composition and employment of air forces. For many years, such debate was largely a theoretical exercise carried out in the pages of military journals because the Treaty of Trianon forbade Hungary to have an air force. When, emboldened by Germany's example, Hungary began rearming openly, it still found itself severely constrained by financial and technological inadequacies. Unlike the major air powers who went to war with forces that generally reflected their theories of aerial employment, Hungary was forced to accept whatever aircraft Italy and Germany were willing to release for export.

1941: Yugoslavia and Early Operations against the Soviet Union

Hungary entered the Second World War on the side of the Axis on 11 April 1941, when its army (the Honvédség) joined the German invasion of Yugoslavia. The proclamation of an independent Croatian state one day earlier offered a suitable justification for Admiral Miklós Horthy, the Hungarian regent, to abrogate the Treaty of Eternal Friendship with Yugoslavia negotiated by his prime minister, Count Pál Teleki. Teleki himself was as dead as the eternal friendship. He committed suicide early in the morning of 3 April after it became apparent that his efforts to keep Hungary out of the coming war had failed. Winston Churchill considered Teleki's death "a sacrifice to absolve himself and his people from guilt in the German attack on Yugoslavia."[1]

There were no self-sacrificial gestures from officials in Budapest when Hungary declared war on the Soviet Union on 26 June 1941. After initially declining to join in Operation Barbarossa, Horthy demanded retaliation following an aerial attack on the northern Hungarian city of Kassa in which thirty-two people were killed. Although there were conflicting reports concerning the identity of the Kassa bombers, the Hungarian General Staff (Vezérkarfőnökség, VKF) reported to Horthy that they were Soviet aircraft. The next day, Prime Minister László Bárdossy, Teleki's successor, made this announcement: "The Royal Hungarian Government concludes that in consequence of these attacks a state of war has come into being between Hungary and the Soviet Union." After being interrupted by cheers, Bárdossy continued, "Only one more sentence. The Hungarian Air Force will take the appropriate measures of reprisal."[2]

Horthy's instrument of revenge, the Royal Hungarian Air Force (Magyar Királyi Honvéd Légierő, MKHL), conducted its retaliatory raid against an airfield and barracks in the Ukrainian town of Stanislau with a mixed force of bombers and escort fighters of Italian and German manufacture. The actual damage inflicted by the MKHL was not precisely determined, but resistance was recorded as heavy, and two Fiat CR.42 fighters failed to return. The VKF was nonetheless pleased: "This morning, the Hungarian Air Force executed very effective reprisal attacks against military targets in Soviet Russia." Two days later a similar MKHL force attacked a Soviet supply depot in Strij and destroyed a key bridge over the Pruth River. Hungarian ground forces entered the war against the Soviet Union on 1 July, and thereafter

MKHL offensive operations were confined largely to supporting Axis ground campaigns. The air strikes in June 1941, though extremely limited in scope and effect, represented an independent air campaign conducted in pursuit of national objectives, and their end, in the words of a VKF report, "concluded the period of air force strategic employment."[3]

This period of strategic employment was vanishingly brief and marked the beginning of a steep decline in the MKHL's independence, prestige, and influence. Its high-water mark in that regard had been reached two years earlier in a short war against Slovakia. In its first day's battle as an independent service, the Hungarian Air Force, whose existence had been publicly declared only in September 1938, had triumphed over a despised enemy, downing eight Slovak fighters without losing any of its own aircraft to enemy action. Admiral Horthy issued a special order to praise the Hungarian airmen, and European newspaper coverage substantiated the MKHL's claims of success.[4] Less than two years later, however, the MKHL would lose its status as an autonomous branch of the Honvéd. The VKF was opposed to the air service's independence, and an accumulation of administrative failures convinced Horthy to place the MKHL once again under the VKF's authority. An infantry general was appointed its head.[5]

The fate of the Hungarian Air Force in these years mirrored that of the country itself. The MKHL had by 1941 already reached its zenith and had been forced to surrender its short-lived autonomy to the army. As the Axis war in Soviet Russia continued, Hungary and its armed forces would gradually be subsumed by the Nazi state. MKHL losses were made good with German aircraft of increasing capability, but with the replacement aircraft came greater Nazi oversight and control, until the MKHL, which entered the war as an ally, would be little more than a Luftwaffe auxiliary unit.

Hungarian airmen were well accustomed to a pronounced German presence. They knew, in fact, little else because throughout the First World War the Austro-Hungarian air service, the Luftfahrtruppen, had been reliant on German aviation technology. Berlin had vastly outspent Vienna in the years leading up to the First World War, and its investment in airframe and engine production carried the Central Powers through the end of the war. Austro-Hungarian aircraft firms manufactured only 10 percent of the aircraft produced by each of the other major combatants.[6] The remainder of the Luftfahrtruppen's needs were met by German imports. The German supply line dried up in the years after the war when the former Central Powers were prohibited by the Paris treaties from fielding air forces, but even then the German influence was strongly felt. István Petróczy, the first Magyar military pilot and leader of Hungary's clandestine postwar air staff, studied the successful German transition from military to civilian aviation and aimed to emulate it.[7] Petróczy's efforts to build Hungary into a self-sustaining regional air power failed. A unified postwar Austria-Hungary, with natural resources around the periphery of the empire and modern manufacturing at its center, might have managed the transition, but Budapest, separated from both by treaty, could not.

German influence in the Hungarian armed forces extended beyond technology into military culture. German was the command and service language of the Habsburg Common Army, of which the Luftfahrtruppen was part, and also the language of Austro-Hungarian military flight instruction. Magyar airmen, therefore, would have been very comfortable flying German airplanes with a German vocabulary, even as they tended to congregate in flying units commanded by fellow Hungarians. In addition, a large percentage of Honvéd officers (particularly at higher ranks) were ethnic Germans whose families had chosen to remain in Hungary after the imperial dissolution and had Magyarized their names. The MKHL's leading air power theorist, Ferenc Szentnémedy (née Willwerth), was an ethnic German, as were four of the five chiefs of staff in the critical period 1938–1945: Lajos Keresztes-Fischer (Fischer), Henrik Werth, Ferenc Szombathelyi (Knausz), and Károly Beregfy (Berger).[8] It does not necessarily follow that these German Hungarian officers were disloyal to Budapest, but it does illuminate the degree to which the Honvéd was habituated to German military practices. These cultural sympathies intensified the professional and ideological rationales and resulted in a definite pro-German orientation within the Hungarian armed forces. This inclination, which Horthy recognized and occasionally tried to counter, encouraged the VKF to advocate joining the Nazi war against Yugoslavia and the Soviet Union.

The broad military-cultural understanding did not mean a seamless integration. Friction was felt immediately as the Germans perceived the Hungarians' lack of urgency in pursuing the retreating Soviets. While MKHL units trickled from Hungary into Ukraine in July 1941, the Red Army rushed eastward ahead of the advancing Axis formations. The pace of the MKHL's advance was a source of frustration for Hungary's liaison officer with the Luftwaffe's 4th Air Fleet, who complained to the VKF: "It is my impression that the Germans would appreciate it if our combat air forces would deploy across the Carpathians and participate in the pursuit of the enemy, in light of the fact that the 4th Air Fleet's range is 1,000 kilometers."[9] The 1st Aviation Brigade increased the pace of deployment in response. Both the need for the forward movement and the difficulties the shifts caused are evident in one MKHL bomber pilot's recollection of this period. It had become hard, he remembered, to find Soviet troop concentrations "because they were retreating so fast away from us. . . . There were days when we had to fly 100 kilometers forward, and we still could not find them. Another disadvantage was that the ground details could not keep up with us. We would move forward from one base to the next—no kitchen, no ground personnel—nothing."[10] The race to cross the Dniester ended in disappointment on 6 July, when retreating Soviet troops destroyed the last of the remaining bridges. In the first ten days of the war, the MKHL flew 145 combat missions and dropped 113 tons of bombs. Its losses were few, but aircrews struggled with the unpredictable weather above the Carpathians, regularly encountering snow showers and icy conditions, for which their Italian aircraft were ill equipped.[11]

During the pause at the Dniester, the Hungarian expeditionary force experienced a number of organizational changes, none of them large but all having the effect of bringing them under closer German control. Chief of Staff General Henrik Werth agreed to split General Ferenc Szombathelyi's Carpathian Group (Szombathelyi would become chief of staff in September; until then he commanded the Honvéd's largest maneuver unit in Russia). Two brigades would remain west of the Dniester as occupation troops, while the 24,000 men of the Rapid Corps would continue to advance with Army Group South.[12] To improve air force coordination with the Honvéd troops, the VKF dissolved the 4th Bomber Regiment and created in its place a temporary Rapid Corps Air Group, consisting of a fighter squadron, two reconnaissance squadrons, air defense artillery batteries, two motorized sections, and a mobile field repair unit. The bomber squadrons remained engaged, but they were under centralized control from Budapest. Their activities became increasingly restricted because of German concerns about the difficulty of identifying allied troops on the chaotic battlefield. Hungarian bombers were first confined to targets within the Rapid Corps' sector, but by the third week of July they were being used only for long-range reconnaissance. These limitations frustrated the Air Group's commander, as did the Rapid Corps' inability to provide adequate supplies for the Air Group, but the VKF rejected the commander's proposal for greater independence.[13] With two bomber squadrons' capabilities badly underutilized and airplanes in need of mechanical attention, MKHL headquarters (Légierő Parancsnokság, or Lepság) withdrew the squadrons on 17 July. The mixed squadron composed of six Italian Ca-135s and nine German Ju 86s remained.[14]

These aircraft executed one of the MKHL's most effective strikes in the Axis advance. On 11 August, they attacked the mile-long bridge across the Bug River at Nikolayev, which was the best means for the 60,000 troops of the Soviet Ninth Army to escape encirclement. The mixed bomber squadron's Ca-135s dropped a bridge span and attacked the railway station, and with the escort fighters (six Fiat CR.42 biplanes and five Reggiano Re-2000s in their MKHL debut) claimed eight I-16 Ratas. General Alexander Löhr, the former Habsburg airman now in command of the Luftwaffe's 4th Air Fleet, personally awarded the Iron Cross to several Hungarian airmen on the Nikolayev raid.[15] This mission and others like it contributed to the success of the 17th Army operation. Nikolayev fell on 16 August, and its capture forced the 9th Army farther east and left Soviet forces isolated in the critical Black Sea port of Odessa.[16]

After the capture of Nikolayev, the Rapid Corps had a period of rest and performed security operations on the 17th Army's right flank. As a consequence, MKHL air activity slowed. In July and August 1941, the Rapid Corps Air Group flew 555 missions, dropped forty tons of munitions, and claimed twenty-seven aerial victories. These actions came at the cost of four aircraft and six fliers lost in action. The Air Group's strength was slowly whittled away as units returned to Hungary and were not replaced. In early October, a reconnaissance squadron was withdrawn, followed soon by the withdrawal of the Re-2000s. The remaining fighter squadron returned to Hungary in late November, after flying more than 800 combat hours and shooting down nineteen Soviet planes at a loss of two CR.42s. By the end of 1941, the Rapid Corps and the entire MKHL expeditionary force were back on Hungarian soil. Fifty-six Hungarian planes had been destroyed or damaged beyond repair, and thirty-six airmen were dead or missing.[17]

1942: AIR SUPPORT FOR THE HUNGARIAN 2ND ARMY

The withdrawal of the Rapid Corps from Ukraine in December 1941 was the result of a debate between Werth and Szombathelyi. As early as August, Werth had begun to arrange training in Germany for the Rapid Corps' reinforcements that he had promised Franz Halder, chief of the German Army High Command. Werth had not informed Horthy of this plan, but he sent Bárdossy a long memorandum taking the Hungarian government to task for its lack of enthusiasm for the war and suggesting that the prime minister offer the Germans at least four additional army corps. Bárdossy passed the substance of Werth's harangue to Horthy and politely told the regent to choose between his prime minister and his chief of staff.[18] Shortly after this, Horthy relieved Werth and named Szombathelyi chief of the VKF. Szombathelyi's first official duty was to accompany Horthy and Bárdossy to Germany for talks with Adolf Hitler, the German foreign minister Joachim von Ribbentrop, and the German General Staff. The regent asked Hitler to release the Honvédség from the eastern front. Ribbentrop countered with the suggestion that a withdrawal would "be very alarming from the point of view of morale" and offered instead to reequip the Rapid Corps' armored brigades.[19] Further, the Germans asked for replacements for the Mountain and Frontier Brigades and an additional twelve battalions, with the promise that these units would be used only for occupation and security duties. Because the Hungarians were eager for the return of their elite Rapid Corps and the offer to refit the brigades seemed reasonable, Bárdossy agreed. He tried to squeeze Ribbentrop for further border revision to the northeast and southwest, but the German was noncommittal. When pressed again on the same issue later, Ribbentrop told Bárdossy, "As long as the war goes on, family quarrels must be suspended."[20]

Germany's attitude toward Hungarian participation in the war changed dramatically after the success of the Soviet counteroffensive in the Moscow sector in December 1941. Just after Christmas, the führer wrote to stiffen Horthy's resolve in their shared fight against Bolshevism and to ask for additional Hungarian forces for the summer offensive. Ribbentrop presented the full proposal in Budapest: Hitler wanted the entire Honvédség put at his disposal. Szombathelyi later described the Germans as modifying their view of Hungary's role from "voluntary participation into an obligation."[21] Ribbentrop used his customary tactic of comparing Hungary's commitment unfavorably to that of Romania and implying that a lack of cooperation on Budapest's part would put at risk the Vienna Awards of 1938 and 1940, which had repatriated to Hungary substantial territory lost in the Paris peace treaty. Bárdossy conceded in principle to a larger Hungarian contingent than agreed upon in September, but both he and Szombathelyi were surprised at the German High

Command's opening demand for twenty-three brigades. Szombathelyi eventually consented to sending nine brigades, with supporting armor and air formations, and to allow a 20,000-man Waffen SS division to be formed from among Hungary's ethnic German population.[22]

Between April and July 1942, the Hungarian 2nd Army, under the command of General Gustáv Jány, moved east in the largest deployment in the country's history. By the first week of June 1942, the 2nd Aviation Brigade, Jány's air component, had assembled in Ukraine. It consisted of two reconnaissance squadrons (with a total of four He 111s and twelve He 46s), two fighter squadrons (each with eleven Re-2000 Héjas), and one bomber squadron (with fourteen Ca-135s). The brigade also had attached six transport aircraft (each with three Ca-101s and converted Ju 86s) and a few Bü 131 and FW 58 trainers for liaison flights.[23] In place of the tricolor chevron adopted in September 1938, the Hungarian aircraft now bore a new marking—a bold white cross inside a black square—consistent with other Axis air force insignia. The white cross was adopted primarily to ease identification of friendly aircraft, but it also served as a visible symbol of the increasing influence of the Luftwaffe over the MKHL. The 2nd Aviation Brigade was the rough equivalent of the Rapid Corps Air Group, fielded the previous year. However, at 200,000 men the 2nd Army was twice the size of the Carpathian Group, and Jány's force would be spread over an area many times larger than the narrow sector through which Szombathelyi advanced.[24] More worryingly, the Red Air Force was far more capable in June 1942 than it had been the previous year. A large proportion of the 5,300 Soviet aircraft destroyed in 1941 were obsolete designs already scheduled to be replaced, meaning "the Luftwaffe simply completed a job that was already being carried out by the Red Air Force itself." Soviet aircraft factories made up the losses by the end of 1941, producing 5,100 Yak-1, MiG-3, and LaGG-3 fighters, each of which compared favorably with the Héja.[25]

Operation Blue, the German summer offensive designed to seize the Caucasus oil fields, began on 28 June 1942. The plan called for Army Group B to capture Voronezh, a city on the east side of the Don River, and then follow the Don south, trapping and eventually destroying the Soviet armies on the west side of the river. The Hungarian, Romanian, and Italian contingents would then be left to guard the Don while the German Army pushed south to the Caucasus.[26] Joseph Stalin was convinced the capture of Moscow would be the main German objective in 1942, and the southern attack took the Red

Army by surprise. By the end of July, the Hungarian 2nd Army, exhausted after slogging more than 600 miles (966 kilometers) across Ukraine, assumed defensive positions along a 120-mile (193-kilometer) section of the Don. They soon had orders to seize three crucial bridges that remained in Soviet hands along the west bank, but owing to the Hungarians' lack of heavy weapons, the Red Army units were able to beat back their attacks.[27] Halder did not appreciate the 2nd Army's efforts, as evidenced by some of his diary entries in August 1942: "The Hungarians are allowing the Russians to come back across the Don! . . . South of Voronezh the Hungarians are running. . . . The Hungarians are making no progress in cleaning up the west bank of the Don. They stop trying and take up defensive positions."[28] This was not an entirely fair assessment of the green Hungarian Army's performance. Because Horthy wanted to minimize Hungary's contribution to the German war effort, he had sent the smallest force he could manage. Each division was short an infantry regiment, so that the nine divisions of the 2nd Army were equal in manpower to only six German divisions. The disparity in weapons was even more striking: the Hungarian formations had half of the light machine guns and anti-tank guns of their German counterparts and only 10 percent of the trucks and tractors.[29] As in the First World War, when compared to a German division, the Hungarian unit was "not only smaller [but] also qualitatively inferior."[30]

The 2nd Aviation Brigade could not compensate for the 2nd Army's lack of heavy artillery and antitank weapons. The brigade's reconnaissance units were able to provide useful observation of Soviet movements behind the bridgeheads, but raids by the 4/1 Bomber Squadron and the Héjas did not break Red Army resistance in the area. To add to the brigade's firepower, He 46s began to conduct armed reconnaissance from early August. Observation crews were allowed to attack some targets on their own if the targets were weakly defended and outside the range of 2nd Army artillery batteries.[31] Giving reconnaissance crews the freedom to attack the enemy rather than just observe and photograph probably increased morale in the unit, but it had no appreciable effect on Soviet strength in the area because the He 46s were still too few and too lightly armed. The MKHL needed heavy, rugged airplanes capable of conducting accurate strikes on small targets and operating with minimal maintenance. This description fit the Ju 87, for instance, which the MKHL would field in 1943, and the excellent Red Air Force Il-2, but not the Ca-135 or the Re-2000. The latter two aircraft were somewhat fragile Italian machines that required significant mechanical attention. Lack of spare parts and battlefield attrition gradually reduced the number of mission-ready aircraft in the brigade, so that in the middle of August only twenty-five aircraft were available for combat.[32] By October 1942, the MKHL's 2nd Aviation Brigade had withdrawn its Caproni bombers entirely, and the MKHL itself took on an increasingly German aspect.

The brigade's long-range reconnaissance squadron, whose He 111s had first been replaced by Dornier Do 215s confiscated from Yugoslavia, received Ju 88s from the Luftwaffe in November. Its 1/1 Fighter Squadron had also begun operating German aircraft because the brigade commander had convinced the Luftwaffe to supply twelve Bf 109s for operational conversion in the field. Thirteen fighter pilots trained by Jagdgeschwader 52 formed the initial Hungarian Bf-109 cadre. After providing the aircraft and training, Jagdgeschwader 52 shared operational control of 1/1 Fighter Squadron with the 2nd Aviation Brigade.[33] Although this arrangement provided Hungarian pilots with front-line German machines and training, it also marked the beginning of the end of the autonomous MKHL.

1943: Disaster on the Don

On 12 January 1943, the Soviet counteroffensive that had begun in November 1942 around Stalingrad reached the Hungarians on the Don River. The Red Air Force committed two air armies to the Voronezh front: 400 combat aircraft, including 138 fighters and 154 attack aircraft.[34] The assault began when two Red Army divisions launched a probing attack from the Uriv bridgehead that penetrated 3 miles (4.8 kilometers) into the Hungarian sector. A counterattack by the German 700th Armored Group on 13 January failed to dislodge the Soviets. On 14 January, the Soviet 40th Army tore a 25-mile-wide (40-kilometer-wide) gap in the 2nd Army's lines.[35] General Gusztáv Jány requested permission to retreat on 15 January, but Army Group B headquarters refused, saying that only Hitler could approve such a request. After permission was denied again the next day, Jány appealed to Szombathelyi, who confirmed the need to follow German orders. With the situation deteriorating quickly, Jány finally ordered retreat on the morning of 17 January.[36]

The 2nd Aviation Brigade's fighter squadrons were based at Ilovskoje, which happened to be the point on which the Soviet pincers were converging. After the Luftwaffe ordered its units there to fall back west, Lieutenant

Colonel Fráter, the brigade commander, asked to follow suit. He was told to wait. With the Red Army approaching on 18 January, the evacuation of Ilovskoje was approved. The transport squadron's Ju 52s and Ju 88s from the reconnaissance squadron carried away as many airmen and supplies as possible. The Re-2000s would not start in the bitter cold (−20 degrees C), and ten Héjas, along with one Ar 96, were blown up lest they fall into Soviet hands. In the battle to defend the airfield on 19 January, a fighter-squadron commander and approximately fifty MKHL personnel were killed. The Ilovskoje disaster forced a stand-down and reorganization of the 2nd Aviation Brigade. All of the 5/2 Fighter Squadron's Re-2000s had been destroyed. The brigade lost a total of thirty-six aircraft and eighty-two airmen in the winter fighting. In addition to the Héjas torched at Ilovskoje, seven Bf 109s, five Ar 96s, and two each of the Ju 88s, Ju 86s, and FW 58s were destroyed.[37]

These losses were nothing in comparison to the catastrophe that befell the rest of the Hungarian 2nd Army. Over the course of the 2nd Army's deployment, it suffered casualties in excess of 50 percent, the vast majority of them coming in January and February 1943. Of the 250,000 Hungarians (including the approximately 50,000 Jewish laborers) who departed for Ukraine in spring 1942, approximately 100,000 were killed or wounded, and 28,000 taken prisoner.[38] For Hungary's future military capacity, the destruction of equipment was even more devastating: "All of the armor was lost; almost all of the artillery (one regiment saved one gun out of twenty-four); 70 to 80 percent of the heavier arms of the infantry (machine guns, trench mortars, etc.), about half the horses, practically all the stores, and a high proportion even of the rifles, for many of the exhausted men had jettisoned even these in their flight."[39] The defense minister estimated that all of the arms left in Hungary could equip only four and a half light divisions. The 2nd Aviation Brigade had only thirty-three airplanes on 1 March 1943.[40] The Honvédség needed German weapons.

Hitler was not, however, inclined to supply them. He had always favored the Romanians in arms shipments, primarily because Romanian oil was dearer to him than Hungarian grain, but he had also come to doubt the Hungarian fighting spirit, at least when the fighting was done on Germany's behalf. When approached by the Hungarians with a request for Bf 109s in 1942, the führer gave this response: "That would just suit the Hungarian gentlemen! They would not use the single-seaters against the enemy but just for pleasure flights! Aviation fuel is in short supply and I need pilots who attack and not ones who go on pleasure flights. What the Hungarians have achieved in the aviation field to date is more than paltry. If I am going to give some aircraft, then rather to the Croats, who have proved they have an offensive spirit. To date we have only experienced fiascos with the Hungarians."[41]

Fortunately for the MKHL, an arrangement had been made that did not rely on Hitler's steadfast goodwill. In the spring of 1941, negotiations between the German and Hungarian Defense Ministries had resulted in a joint air-craft-production program for the manufacture of Bf 109s, Me 210s, and Daimler-Benz DB-601 engines, from which the MKHL anticipated obtaining 225 Bf 109s and 160 Me 210s. With the exception of Reggiane and Caproni deliveries on prewar contracts, there had been no substantial deliveries of combat aircraft to Hungary from foreign producers after the invasion of the Soviet Union.[42] The Regia Aeronautica and Luftwaffe needed every up-to-date combat airplane that their factories could construct, leaving only train-ers and obsolete models for purchase. There were direct transfers of Luftwaffe aircraft to the MKHL, however, such as the one that gave twelve Bf-109s to the 1/1 Fighter Squadron in early 1943. Between June 1942 and March 1944, at least 170 of the MKHL's combat aircraft were passed on directly from Luft-waffe units.[43] It is possible that the number of aircraft moved from German to Hungarian units off the books was substantially higher because the 400 Bf 109s that the 101st Fighter Regiment reported fielding in 1944 nearly equaled the total amount of Hungarian production (488).[44]

The transfer of aircraft in the field was undoubtedly appreciated by the local MKHL commanders, but these aircraft were considered favors granted, not contracts fulfilled, and therefore often resulted in the unit receiving the Luftwaffe's cast-offs and coming under some measure of German command. That was indeed the situation for Hungarian airmen in the East for most of 1943. The 2nd Aviation Brigade still existed as an organization, but it remained extremely weak. There was little will in Budapest to reinforce it, and until production ramped up late in the year, there were no suitable aircraft to send. Fighter pilots of the 1/1 and 5/2 Squadrons flew Bf 109s in Ukraine, mostly conducting ground attack sorties with 500-pound bombs and occasionally escorting Luftwaffe bombers. They scored twenty aerial victories during the battle of Kursk and accounted for 132 Soviet aircraft destroyed by the end of 1943. Through the summer, the Hungarian units retreated west with the Luft-waffe, passing back along the route they had followed two years earlier.[45] After their recall from the front in the autumn of 1942, 3/1 and 4/1 Bomber

Squadron pilots departed for Istres, France, where they learned to fly Ju 88s, the Luftwaffe's standard light bomber. The 1/1 Long-Range Reconnaissance Squadron had by that time also converted to Ju 88s, although without the benefit of a stay in France, and it operated wholly under the command of the Luftwaffe.[46]

In January 1944, the MKHL reorganized to focus on the aerial defense of Hungary. Its expeditionary force in the Soviet Union had shrunk to fewer than thirty aircraft. Security of the capital was trusted to three fighter squadrons equipped with Héjas, Bf 109s, and Me 210s (for night interceptions). The border defense group comprised thirteen squadrons, flying an assortment of aircraft, from Ca-135s and He 46s procured in the late 1930s to Bf 109s fresh from the factory.[47]

Even as the MKHL was taking steps to strengthen capital defense, the government was desperate to exit the war. Horthy's determination to secure a separate peace had intensified after the destruction of the 2nd Army, and Hungarian peace feelers had gone out all over Europe. Horthy had sacked Prime Minister Bárdossy in March 1942, in part over Bárdossy's hasty declaration of war against the United States, and had charged his replacement, Miklós Kállay, to "regain Hungary's freedom of action and return, if possible, to a state of non-belligerence."[48] Kállay's emissaries were authorized to inform British or American officials that Hungary would not resist an Anglo-American or Polish army and that the Honvédség was prepared to act against the Wehrmacht if required. Hitler was well aware of Kállay's pro-Western orientation and had confronted Horthy about it in April 1943. Horthy defended his premier, but his protestations did nothing to assuage German suspicion. Ribbentrop remarked to the German minister in Budapest that Horthy's support of Kállay only proved "the regent's full approval of the prime minister's policy of defeatism and detachment from the Axis Powers."[49] After Italy defected to the Allies in September 1943, Hitler ordered that plans be prepared for the occupation of Hungary and Romania.[50]

1944 TO THE WAR'S END: GERMAN OCCUPATION, AMERICAN BOMBING, AND THE SOVIET OFFENSIVE

Operation Margarethe I was implemented on 19 March 1944. Two days earlier Hitler had called Horthy to Klessheim Castle to coerce him into agreeing to German occupation. Hitler demanded the reinstatement of senior pro-German officials in the Hungarian government, and Horthy was to instruct the Hungarian people to welcome the occupying Wehrmacht. The regent refused. After many hours, a compromise emerged. Horthy would appoint an acceptable government, but German troops must thereafter be withdrawn.[51] The German military attaché informed Kállay that paratroops would occupy airfields around the capital at 4:00 a.m. and that eleven German divisions would arrive in Budapest by 6:00. The VKF estimated nine or ten German divisions were poised on Hungary's borders, along with ten Romanian divisions and one Slovak. The only Honvéd units capable of sustained resistance were in the Carpathians. Kállay ordered the MKHL to disperse available aircraft to avoid possible internment by German paratroops. The entire government resigned the next day. Döme Sztójay became prime minister. Even though Sztójay was one of Hitler's favorite Hungarians, Horthy believed that as a soldier Sztójay would prove loyal to the regent as head of the armed forces. Horthy went into semi-exile in the palace.[52]

In the first days of the occupation, the Wehrmacht confined Honvédség troops to the garrisons. Although the operational plan involved disarming the Hungarian soldiers, it was not widely implemented. The Honvédség was loyal to the regent, who had, through the VHK chief, instructed the army to admit German forces to their installations. Subsequent actions showed the Germans had no confidence in the Honvéd's loyalty to the Reich. General Hans von Greiffenberg, promoted after the occupation from attaché to "Plenipotentiary General of the Wehrmacht in Hungary," established German military control in eastern Hungary and prepared for a purge of the officer corps. Honvéd generals and political leaders known to be hostile to Nazism were arrested.[53]

The German occupation had two effects on the MKHL. The immediate effect was that all units were grounded until the situation had settled. When it was apparent that there would be no defections, normal operations resumed. The definition of normal operations changed right away, however, because in the aftermath of the occupation Allied bombers began striking targets in Hungary. Before this point, as long as there had been the possibility of a separate peace with Budapest, the B-17s and B-24s of the US 15th Air Force had previously passed over Hungarian airspace with little resistance on their way to targets in Austria and Romania.[54]

That uneasy existence ended on 3 April with a 200-bomber raid on Hungarian aircraft factories. The air raid sirens sounding in Budapest were mostly ignored because there had not been an aerial attack on the capital since a pair of ineffectual Soviet raids in September 1942.[55] Through the month of April, the raids were sporadic but destructive. The MKHL responded to each one, but with scant results, in part due to the deficiencies of the air defense system. Poor communications and inexperience meant the ground controllers sometimes vectored the fighters to an empty piece of sky. On 13 April, they compounded the error by directing the fighters to land to refuel just as bombs started landing on their airfield.[56]

Attacks intensified through the spring and early summer. The planners of the Anglo-American Combined Bomber Offensive were instructed to give "top priority to bombing communications in Romanian and Hungary and to treat the whole European transport system 'as one' when undermining German mobility." Beginning in June 1944, the Allies' concentration on destroying the sources of German oil also brought more bombers to Hungary, whose oil fields south of Lake Balaton had grown from producing almost nothing before the war to an annual production of 840,000 tons per year.[57] Hundreds of Allied bombers attacked in waves, and the defenders, despite being reinforced by eighty fighters of the Luftwaffe's 8th Fighter Wing, were simply overwhelmed. The 14 June attack on the Pét nitrogen plants was a red-letter day for the combined Axis air defense forces: they claimed eighteen American aircraft destroyed (eleven to antiaircraft artillery, five to MKHL fighters, and two to the Luftwaffe). That success was followed by the "black day" raids on 16 June. Twenty-eight MKHL fighters scrambled to meet the American force of 658 bombers and 290 fighters. Five Allied P-38s fell, but at the astounding cost of thirteen Hungarian fighters.[58] Not all raids were so one-sided, but even when the MKHL defenders exacted a heavy cost from the attackers, as in the Pét raid on 14 June, it made no difference to the direction of the battle. The Americans came back the next day or the next week with bigger formations, while the Hungarian force just got smaller. By September, fuel shortages kept MKHL fighters on the ground during many US Army Air Forces bombing raids. As refined aviation fuel became scarcer, the Luftwaffe retained most of the stock for its own use.[59] With fewer fighters challenging the bomber force, American fighter aircraft began low-level attacks on Hungarian airfields, which proved effective in further eroding the MKHL's air defense fleet.

That fleet had in fact performed well in comparison to its German counterparts. According to Luftwaffe records, from March to November 1944 the height of the Allied bombing raids, its pilots flew 932 daytime air defense sorties over Hungary. Those flights resulted in seventy-three confirmed victories against American attackers, at a loss of eighty-eight Luftwaffe airplanes and forty-three casualties. In the same time, the MKHL flew 649 defensive missions and scored 107 kills, while losing seventy-eight aircraft and thirty pilots.[60] These figures show that the Hungarians destroyed an enemy aircraft for every six sorties flown, whereas the Germans needed thirteen flights on average to achieve a victory. The MKHL suffered a higher rate of material losses (one per eight flights) than the Luftwaffe (one every eleven flights), but pilot casualties were identical (one loss per twenty-two sorties). The motivation of defending one's homeland could account for the MKHL's greater lethality, as could the decreasing experience level of German pilots owing to the extremely high casualty rates in other theaters.[61]

In early October 1944, the Red Army began its attack on Debrecen, and one month later the Hungarian capital defense fighters were sent east to try to stem the Soviet advance. By that time, the Royal Hungarian Air Force had ceased to exist in any significant way because Hungary itself had become a German puppet state. In the summer, as German losses mounted, Horthy attempted to reassert himself as the head of state. After bringing an end to the deportation of Hungary's Jews to the death camps, Horthy forced Sztójay to resign. He tried to open negotiations with the Allies through personal emissaries. Western leaders were not receptive, but acceptable terms were reached with the Soviets on 10 October. According to those terms, Hungary would cease operations against the Soviet Union, declare war against Germany, and within ten days pull its forces out of territory gained after 1937. The Germans caught wind of the negotiations, of course, and German commando Major Otto Skorzeny kidnapped Horthy's surviving son (his older son and vice regent had been killed in a Héja crash in Russia in 1942) and held him at Mauthausen. Horthy was informed of this development just before making the national proclamation of Hungary's defection from the Axis. His message was less than completely clear, and very few Honvédség formations changed sides. Horthy was stunned. The German Army was not and fully occupied Budapest overnight on 15 October 1944. The regent and his remaining family members were sent into house arrest in Bavaria.[62]

The Hungarian Army, of whose loyalty Horthy had been confident, failed almost entirely to heed his order on 15 October. General Béla Miklós, 1st Army commander, directed his senior subordinate commanders to obey the regent and defect. Not a single corps or division commander did so. One regimental commander who attempted to change sides had his own order disobeyed and was executed by the Germans.[63] Each commander who continued to fight alongside the Wehrmacht would have had his own set of rationales for his actions, including an ideological affinity for National Socialism, a strong pro-German ethnic or cultural identity, antipathy toward communism, fear of foreign occupation, desire to retain the Vienna Award territories, feelings of camaraderie for German soldiers, and simple confusion about the validity of the Horthy order. It is of course possible that all of those notions could be held at once without serious cognitive dissonance and without feeling oneself a traitor to Hungary. Certainly there were officers who were so committed to the tenets of National Socialism that their loyalty to Hungary came a poor second. But even for those Honvédség officers who were not ideologically motivated to cling to Germany, the idea of swapping sides was distasteful. It must have been nearly impossible for them to see the advancing Red Army as a potential ally and the Germans they had fought alongside for three years as the enemy. For twenty-five years—or the entire lifetime of a Honvédség company commander—Hungary had mourned its losses in the First World War and reviled the Russian-inspired communism that had preceded the Romanian intervention of 1919. That the order to surrender to the Soviets came from Admiral Horthy, the exemplar of revisionism and antibolshevism, would have seemed too bizarre to credit. The right thing for the nation, these officers might have reasoned, would be to fight the invading Russians, Romanians, and Slovaks for as long as possible.

There were no unit defections in the MKHL. Hungarian airmen would have felt even closer to their German comrades than did soldiers. The relationship between the MKHL and Luftwaffe had been more intimate than that between the Honvédség and the Heer. Hungarian pilots flew German planes with German training and tactics. Although the squadrons themselves remained separate, German and Hungarian pilots often shared the same airfields. From October 1944, the air defense units were subordinated to the Luftwaffe's 8th Fliegerdivision.[64] Hungarian and German pilots also shared a bond as combat airmen. These factors made it extremely unlikely that a large collection of airmen would cross over to the Soviet side. The putsch of 15 October made little operational difference to the MKHL, which was by then a force in unalterable decline, unable to exercise initiative.

Conclusion

By the end of the war, the Royal Hungarian Air Force was scarcely distinguishable from the Luftwaffe. The MKHL began the war as an allied force and ended it as an auxiliary. From the perspective of the senior partner, this was a satisfactory arrangement. Hitler's fears notwithstanding, the "Hungarian gentlemen" did not use the fighters supplied by Berlin for pleasure flights but rather stayed engaged against the Allies even after Hungary was overrun by the Red Army. The MKHL provided a stock, although admittedly small, of the one critical resource for which the Luftwaffe could find no substitute: trained fighter pilots. The Hungarian air defense force was by late 1944 completely subordinated to the Luftwaffe for matters of command and logistics, retaining its own authority only for personnel matters.

From the viewpoint of the junior partner, the alliance with Germany was more problematic. Increasing German domination in the relationship meant, on the one hand, the MKHL was equipped with front-line combat aircraft but also signaled, on the other, further erosion of the MKHL's autonomy. That autonomy had been in steady decline since the high-water mark of 1939 and in some sense had always been illusory. Hungary's economic and industrial weakness limited its ability to produce aircraft, and its diplomatic isolation and financial instability meant that most foreign manufacturers were beyond its reach. Only through arms credits, first from Benito Mussolini and later and most importantly from Hitler, could the MKHL acquire modern aircraft types in any substantial numbers. Even with the Italian and German credits, the Hungarians were forced to accept aircraft that the Regia Aeronautica and Luftwaffe had found undesirable or that had been produced in surplus numbers. Not until Axis production began to outstrip the Luftwaffe's employment capacity were Hungarian pilots equipped with top-shelf aircraft. Thus, the most important area of German–Hungarian aerial cooperation was material. The MKHL leaders understood how to train pilots and had a conception of air power employment informed by the robust interwar debates that had swirled through Western air forces. Hungary could not, however, build enough modern aircraft to equip itself and therefore was completely reliant on its allies to supply its fleet.

This problem came as no surprise. Kálmán Csukás, who later became a fighter squadron commander and was killed in Russia, recognized in 1940 the difficulties that a small country would face in trying to conduct an independent air war. He suggested that in the future small countries would probably fight as part of a larger alliance. Further, he wrote, "small countries normally do not possess their own independent aviation industry but rather purchase particular types from large powers. They have to be familiar with the large power's strategic concept because only then can one judge whether the material is suitable for its objective."[65] In the event, the Luftwaffe's operational art matched the MKHL's own needs, but Hitler's strategic concept— unrestricted war in the East with the goal of destroying the Soviet Union—did not suit Hungary at all. Hungarian revisionism, even in its maximalist form, had fairly limited aims: it required only the reoccupation of contiguous territories sliced away at Trianon. The combined German–Hungarian air arms were well configured to achieve Budapest's modest territorial aims but were wholly inadequate for Berlin's objective of continental dominance.

Notes

1. Quoted in Bryan Cartledge, *The Will to Survive: A History of Hungary* (London: Hurst, 2006), 381.

2. Quoted in C. A. Macartney, *October Fifteenth: A History of Modern Hungary, 1929–1945*, part 2 (Edinburgh: Edinburgh University Press, 1956), 29–30.

3. Quoted in Janós Vesztényi, "A magyar katonai repülés, 1920–1945," 1978, 2.787, V/255–56, Hadtörténelmi Levéltár Tanulmánygyűjtemény, Budapest. All translations are mine unless otherwise noted.

4. Csaba Stenge, *Baptism of Fire: The First Combat Experiences of the Royal Hungarian Air Force and Slovak Air Force, March 1939* (Solihull, UK: Helion, 2013), 60–66, 84.

5. M. Miklós Szabó, *A Magyar Királyi Honvéd Légierő, 1938–1945: Elméleti-technikai-szervezeti fejlődése és háborús alkalmazása* (Budapest: Zrínyi Kiadó, 1999), 96.

6. John Ellis and Michael Cox, *The World War I Databook: The Essential Facts and Figures for All the Combatants* (London: Aurum Press, 2001), 245, 287.

7. István Petróczy, "A legyőzött Németország aviatikája: Tanulságok és teendők" (1921), 5271, I/3, Hadtörténelmi Intézet Könyvtára, Budapest.

8. ¹ Sándor Szakály, *A magyar katonai elit, 1938–1945* (Budapest: Magvető Konyvkiadó, 1987), 20–52.

9. ¹ Quoted in Szabó, *A Magyar Királyi Honvéd Légierő, 1938–1945*, 166.

10. Quoted in L. Hangodi, "A M. Kir. Honvéd 4/I-es Nehézbombázó-Osztály története, 1936–1945," *Hadtörténelmi Közlemények* 116, no. 1 (2003): 158.

11. György Punka and Gyula Sárhidai, *Magyar sasok: A Magyar Királyi Honvéd Légierő 1920–1945* (Budapest: Zrínyi Kiadó, 2006), 39.

12. Lóránd Dombrády and Sándor Tóth, *A Magyar Királyi Honvédség* (Budapest: Zrínyi Kiadó, 1987), 195–96.

13. Vesztényi, "A magyar katonai repülés," V/269–74.

14. Szabó, *A Magyar Királyi Honvéd Légierő, 1938–1945*, 146–48.

15. J. R. Gaal, "The Bridge over the River Bug," *Air Combat* 6, no. 2 (1978): 80–87.

16. John Erickson, *The Road to Stalingrad* (London: Weidenfeld & Nicolson, 1975), 204.

17. Punka and Sárhidai, *Magyar sasok,* 33, 39.

18. Miklós Horthy, *Horthy Miklós titkos iratai,* ed. Miklós Szinai and László Szűcs (Budapest: Kossuth Könyvkiadó, 1965), 307–8.

19. Macartney, *October Fifteenth,* part 2, 54.

20. The details of this meeting come from Gyula Juhász, *Hungarian Foreign Policy,* trans. Sándor Simon and Mária Kovács (Budapest: Académiai Kiadó, 1979), 200–201.

21. Quoted in Deborah S. Cornelius, *Hungary in World War II: Caught in the Cauldron* (New York: Fordham University Press, 2011), 184.

22. Dombrády and Tóth, *A Magyar Királyi Honvédség,* 226–27. The requested contribution is also given as twenty-six divisions (Szabó) and twelve brigades (Cornelius) (Juhász, *Hungarian Foreign Policy,* 207).

23. Szabó, *A Magyar Királyi Honvéd Légierő, 1938–1945,* 158–59.

24. Cornelius, *Hungary in World War II,* 207–9.

25. Richard J. Overy, *The Air War 1939–1945* (New York: Stein and Day, 1981), 62–63.

26. Gerhard L. Weinberg, *A World at Arms: A Global History of World War II* (Cambridge: Cambridge University Press, 1994), 413–14.

27. Cornelius, *Hungary in World War II,* 208.

28. Quoted in Mario D. Fenyo, *Hitler, Horthy, and Hungary: German–Hungarian Relations, 1941–1944* (New Haven, CT: Yale University Press, 1972), 42.

29. M. Miklós Szabó, *A Magyar Királyi Honvéd Légierő a második világ háborúban* (Budapest: Zrínyi Kiadó, 1987), 117.

30. Hew Strachan, *To Arms,* vol. 1 of *The First World War* (Oxford: Oxford University Press, 2001), 284.

31. Csaba Horváth, "A Magyar 2. Hadsereg Közelfelderítő Repülőszázadának működési rendszere a szabályzatok tükrében 1942 június-október," *Hadtörténelmi Közlemények* 107, no. 2 (1994): 117.

32. Ibid. 123; Szabó, *A Magyar Királyi Honvéd Légierő, 1938–1945,* 187. Twenty-five was the average of seven days' serviceability rates, 11–18 August 1942.

33. Sándor Nagyváradi, M. Miklós Szabó, and László Winkler, *Fejezetek a magyar katonai repülés történetéből* (Budapest: Műszaki Könyvkiadó, 1986), 255–56.

34. Szabó, *A Magyar Királyi Honvéd Légierő, 1938–1945,* 214.

35. Péter Szabó, "Hungarian Soldiers in World War Two: 1941–1945," in *A Millennium of Hungarian Military History,* ed. László Veszprémy and Béla K. Királyi, trans. Eleonóra Arató (New York: Columbia University Press, 2002), 452–54.

36. Cornelius, *Hungary in World War II,* 221–22.

37. Punka and Sárhidai, *Magyar sasok,* 42.

38. P. Szabó, "Hungarian Soldiers in World War Two," 455.

39. Macartney, *October Fifteenth,* part 2, 135.

40. Szabó, *A Magyar Királyi Honvéd Légierő, 1938–1945,* 223.

41. Quoted in Hans Werner Neulen, *In the Skies of Europe: Air Forces Allied to the Luftwaffe 1939–1945,* trans. Alex Vanags-Baginskis (Norfolk, UK: Crowood Press, 2005), 131.

42. Szabó, *A Magyar Királyi Honvéd Légierő, 1938–1945,* 40, 59.

43. Neulen, *In the Skies of Europe*, 131.

44. Punka and Sárhidai, *Magyar sasok*, 52.

45. Neulen, *In the Skies of Europe*, 132.

46. Punka and Sárhidai, *Magyar sasok*, 46.

47. Szabó, *A Magyar Királyi Honvéd Légierő, 1938–1945*, 112.

48. Miklós Horthy, *Admiral Nicholas Horthy: Memoirs,* annotated by Andrew L. Simon (New York: Speller, 1957), 193, 203.

49. Quoted in Juhász, *Hungarian Foreign Policy*, 239.

50. István Mócsy, "Hungary's Failed Strategic Surrender: Secret Wartime Negotiations with Britain," in *Hungary in the Age of Total War (1938–1948),* ed. Nándor Dreisziger (New York: Columbia University Press, 1998), 99.

51. Cornelius, *Hungary in World War II*, 275–78; Juhász, *Hungarian Foreign Policy,* 287–91; Macartney, *October Fifteenth,* part 2, 233–46.

52. Nicholas Kállay, *Hungarian Premier: A Personal Account of a Nation's Struggle in the Second World War* (New York: Columbia University Press, 1954), 419–25, 427–38.

53. Macartney, *October Fifteenth,* part 2, 253–58.

54. J. R. Gaal, "The Bombs of April," *Air Combat* 7, no. 5 (1979): 75.

55. Richard J. Overy, *The Bombing War: Europe 1939–1945* (London: Allen Lane, 2013), 231.

56. Gaal, "The Bombs of April," 77–78.

57. Overy, *The Bombing War,* 293–94.

58. Nagyváradi, Szabó, and Winkler, *Fejezetek a magyar katonai repülés történetéből,* 214.

59. Dénes Bernád, *A nemzet szárnyai: A magyar katonai repülés évszázados története* (Budapest: Zrínyi Kiadó, 2013), 47.

60. Cited in Bernád, *A nemzet szárnyai,* 47.

61. The average Luftwaffe monthly loss rates in all theaters in 1944 was 1,754 aircraft (Overy, *The Air War,* 186). The loss rate over Hungary of five per month from March to November 1944 equals 0.29 percent of that total.

62. Macartney, *October Fifteenth,* part 2, 414–16; Horthy, *Memoirs,* 222–26; Cornelius, *Hungary in World War II,* 312–19.

63. Macartney, *October Fifteenth,* part 2, 445.

64. Neulen, *In the Skies of Europe,* 143.

65. Kálmán Csukás, "Az önálló légiháború," *Magyar Katonai Szemle* 3 (1940): 706.

The Korean War

United Nations Carrier Operations, 1950–1953

Corbin Williamson

Moments after the surrender documents ending the Second World War were signed on Sunday, 2 September 1945, hundreds of American naval aircraft overflew the fleet gathered in Tokyo Bay.[1] This display highlighted the prominent role that carrier-based aviation played in the war against Japan. US Navy (USN) aircraft carriers provided the mobile striking forces used to project American power deep into the western Pacific. In contrast, American carriers in the Korean War were far less mobile, serving more as floating airfields, much as they had been during the invasion of Okinawa in the summer of 1945.[2]

Britain's Royal Navy (RN) was also represented in the Allied armada that accepted Japan's surrender. The RN's British Pacific Fleet had contributed four fleet carriers to the final operations against Japan, although the British force was roughly the size of only one of the four USN task groups composing the US Pacific Fleet's fast carrier striking force.[3] In the Korean War, the USN also provided the preponderance of the United Nations (UN) naval effort, though Britain and the British Commonwealth made important contributions as well.

The experience of British, Australian, and American carrier aviation in the Korean War highlights the value of combined exercises in peacetime, including the natural tensions that arise when services with different organizations, histories, and structures are required to work closely together.[4] Although these differences occasionally caused problems, these issues were overcome through personal relationships at the working level and ad hoc solutions developed in theater. Multinational carrier warfare in Korea benefitted from the combination of the Second World War experience and the postwar efforts to retain interoperability, both of which created trust between the navies.[5]

BRITISH AND AMERICAN NAVAL AVIATION, 1945–1950

In the aftermath of the Second World War, both the British and American navies demobilized, dramatically reducing the number of ships and personnel in their respective services. Britain had spent 28 percent of the nation's total wealth defeating the Axis powers, and Prime Minister Clement Attlee was determined to cut military spending. Attlee ordered major cuts to the personnel strength of each of the services in early 1946. He wanted to reduce the RN down to 175,000 personnel by December 1946, compared to more than 780,000 in mid-1945. Subsequent discussions raised this figure slightly, but the fleet's size fell dramatically, as shown in table 7.1.[6]

By 1946, these cuts reduced the Fleet Air Arm, the naval aviation branch of the RN, to a mere 122 operational planes, roughly one-tenth of its strength in August 1945. In 1947, further cuts left British naval aviation with a handful of light fleet carriers in active service. During the next year, manpower shortages immobilized almost all of the fleet's carriers. The light fleet carriers of the *Majestic* and *Colossus* classes and their embarked squadrons would provide the RN's naval air contribution to the Korean War. However, the reductions in the RN's size meant that in June 1950 the Admiralty maintained only a limited presence in the western Pacific given the fleet's more pressing commitments in European waters. Britain prioritized the Cold War against the Soviet Union in Europe over affairs in East Asia throughout the Korean War.[7]

From the Admiralty's perspective, the RN's principal role in any future conflict was the defense of sea lines of communication against enemy submarines, aircraft, and surface ships. Naval aviation would provide aerial defense of convoys and fleets, wage antisubmarine warfare, and strike enemy ships and shore installations. However, Britain lacked the finances to acquire and maintain in peacetime the number of squadrons required to perform all these roles. The Admiralty accordingly focused on research and development, which produced a number of important technological innovations for carrier aviation: the angled flight deck, the steam catapult, and the mirror landing system.[8]

Table 7.1.   Royal Navy Active-Duty Strength, 1944–1950

Year	Personnel Strength
1944	790,000
1945	788,800
1946	350,000
1947	189,600
1948	135,300
1949	136,900
1950	129,400

Note: Personnel strength listed as of 30 June each year.

Source: UK Central Statistical Office, *Annual Abstract of Statistics No. 90* (London: Her Majesty's Stationery Office, 1953), 95.

In the United States, economic concerns also led to major reductions in military spending. President Harry Truman wanted to reduce government expenditures over concerns about a renewed depression caused by the burden of Second World War debt. Truman thus insisted on major cuts that reduced defense outlays from more than $80 billion in 1945 to $12 billion by 1947, as shown in table 7.2.[9]

At the same time that the US military's budget was falling, President Truman sought to unify the War and Navy Departments into a single Department of Defense. Truman and other officials in Washington felt that the Second World War showed the need for greater cooperation between the services. Truman also wanted to separate the Army Air Forces from the army, creating a separate, third military service. Unification of the services into a single department and the establishment of a separate air force faced strong opposition from officials within the Navy Department.[10]

USN admirals feared that in a unified Department of Defense, army and air force interests would combine to subordinate naval concerns. Furthermore, they also worried that creation of a separate air force might involve the navy losing control of naval aviation to the air force. The RN had lost control of the Royal Naval Air Service when the Royal Air Force (RAF) was created in 1918, leading to decades of tension between the two services. Many American observers felt that RAF control of naval aviation in the 1920s and 1930s had negatively affected British naval aviation's performance in the Second World War and so were not interested in following Britain's model. Shortly

Table 7.2.   US Defense Spending, 1944–1952

Year	Budget*
1944	79,143
1945	82,965
1946	42,681
1947	12,808
1948	9,105
1949	13,150
1950	13,724
1951	23,566
1952	46,089

* In millions of current dollars.

Source: "Historical Tables," in *Budget of the United States Government, Fiscal Year 2019* (Washington, DC: US Government Publication Office, 2018), table 6-1, 122, at https://www.gpo.gov/fdsys/pkg/BUDGET-2019-TAB/pdf/BUDGET-2019-TAB.pdf.

before retiring as the professional head of the RN, Admiral Sir Andrew Cunningham met with Admiral William Leahy, chief of staff to President Truman, on 21 May 1946. Leahy recorded in his diary that Cunningham, the most celebrated British naval officer of the Second World War, "said the British Navy is a horrible example to everybody of the danger of letting [the] Air Force get control of naval aviation."[11]

The US National Security Act of 1947 established a new Department of Defense and a separate Department of the Air Force. The navy retained control of naval aviation, but interservice fighting over roles continued, specifically whether the navy or air force should have control of delivering an atomic bomb. A number of admirals attacked the capabilities of the air force's new B-36 strategic bomber out of concerns for the future of naval aviation. The ensuing debate consumed most of 1949 and led to the firing of the chief of naval operations, Admiral Louis Denfeld, in what became known as the "Revolt of the Admirals."[12] The tensions from this debate within the US military shaped navy–air force relations in the Korean War.

The Korean War Begins

Following Japan's defeat in 1945, the United States and the Soviet Union jointly occupied Korea, dividing the country along the line of thirty-eight degrees north latitude, the thirty-eighth parallel. On the Korean Peninsula, two different visions of the future of Korea developed, a nationalist movement concentrated in the South and a Communist movement centered in the North. In April 1948, a Communist-led insurgency broke out in the South, which the Republic of Korea, established later that August, spent the next two years gradually quelling.

South Korea's internal security campaign, combined with increasing economic and diplomatic links between South Korea and Japan, led Kim Il-Sung, the head of the Democratic People's Republic of Korea, to seek support for a conventional invasion of South Korea. He finally secured Soviet leader Joseph Stalin's support for such an invasion in the spring of 1950 after a series of rejections. By this time, American troops had been withdrawn from South Korea, Mao Zedong's Chinese Communists had evicted their Nationalist opponents from mainland China, and the Soviet Union had tested its first atomic bomb. All these factors made an invasion appear more attractive to Stalin, who sought a foreign-policy success in Northeast Asia after a series of setbacks in Europe. The formation of the North Atlantic Treaty Organization (NATO) in April 1949, the end of the Berlin Blockade in May 1949, and the growing positive impact of the economic aid provided by the Marshall Plan encouraged Stalin to see Kim's proposal in a new light. However, Stalin did not want to be drawn into a direct confrontation with the United States. Therefore, at Stalin's insistence Kim obtained assurances of support from Mao Zedong, chairman of the recently founded People's Republic of China (PRC). North Korean forces accordingly invaded South Korea on 25 June 1950, inflicting heavy casualties on South Korean units, capturing Seoul within days, and making significant territorial gains.[13]

In Washington, the Truman administration's foreign-policy resources in the summer of 1950 were focused on western Europe. In the western Pacific, American attention centered on the occupation of Japan. However, President Truman and his advisers viewed the North Korean invasion as the first step in a larger campaign directed by Moscow that required an American response. His administration, after being blamed by its political opponents for the Communist victory over the Nationalists in China in the fall of 1949, could not afford to lose another US-backed regime to communism. On 27 June, Truman ordered American air and naval forces into action to stem the North Korean offensive, the same day the UN authorized member states to commit military forces to aid South Korea. At the time, the Soviet Union was boycotting the UN Security Council in protest over that body's refusal to give the PRC the Chinese seat on the council currently held by the Nationalists. Truman quickly authorized General Douglas MacArthur, supreme commander for the Allied Powers in Japan and head of Far East Command, to send US ground forces into battle, while Washington began looking for additional nations to contribute forces to the worsening situation.[14]

At the time of the invasion, the USN had two commands in the western Pacific. The 7th Fleet under Vice Admiral Arthur Struble reported to Pearl Harbor, while US Naval Forces Far East under Vice Admiral Turner Joy reported to General MacArthur. Ships from both commands sailed for Korean waters within days of the invasion. On 28 June, the British Admiralty ordered Rear Admiral William Andrewes, RN, flag officer, second in command, Far East Station, to place his small squadron at American disposal in accordance with London's decision to support South Korea. Andrewes had already concentrated the light fleet carrier HMS Triumph along with several escorts in southern Japanese waters and now sailed to join the USN's Task Force 77. Rear Admiral John Hoskins, USN, the task force commander, controlled the fleet carrier USS Valley Forge as well as escorting cruisers and destroyers. Andrewes joined Hoskin's command, and the Anglo-American force sailed for Korean waters. The British adopted American communication procedures and signals, which, according to Andrewes, "seemed so familiar" due to combined exercises held earlier that year.[15]

The two navies had come together in early March 1950 for a series of exercises in the South China Sea. Andrewes with Triumph joined up with Rear Admiral Walter Boone, USN, on the fleet carrier USS Boxer. The British practiced using USN communications and gradually became more proficient with this different signals system. The two carriers also exchanged aircraft, a practice known as "cross-decking." In order to facilitate this interchange, each carrier took on board a landing signal officer to direct carrier aircraft on their final approach. At the time, each navy used different signals to guide aircraft down to the deck. Furthermore, in the RN the landing signal officer's signals were considered orders, whereas in the USN they were considered information for the pilot's use. Due to these differences, aircraft from each carrier operated in single-nation formations and attacked separate practice targets in the Philippines. Reports from both navies agreed that Boxer's flight operations reflected more frequent flying. A British summary analyzing air operations accurately noted that although both navies had officially adopted similar

Senior British and American naval officers participate in a conference on the USS *Rochester*, 1 July 1950. Those present are (*left to right*) Captain Arthur D. Torlesse, RN, commanding officer, HMS *Triumph*; Rear Admiral John Hoskins, USN, commander, Carrier Group, Seventh Fleet; Vice Admiral Arthur D. Struble, USN, commander, Seventh Fleet; and Rear Admiral Sir William G. Andrewes, RN, flag officer, second in command, Far East Station. Photograph 80-G-416423, National Archives at College Park, College Park, MD.

carrier procedures, techniques in practice tended to "drift apart, unless constantly practiced in combined exercises." The exercises showed that carriers from both navies could operate together with little warning and provided the British invaluable practice using American communications.[16]

The March exercises helped RN personnel assimilate rapidly into the USN's Task Force 77 after the North Korean invasion began. On 3 July, *Valley Forge* and *Triumph* launched air strikes against North Korean airfields and bridges near Pyongyang and Haeju. Two days later the two carriers retired to Okinawa for replenishment before heading back to Korean waters on 16 July. Both ships needed to take on fresh supplies, ammunition, and aviation gas before a second round of strikes. Neither navy was fully prepared for extended combat operations in early July, but the carriers' proximity to the Korean Peninsula allowed them to respond rapidly once the invasion began.[17]

These initial strikes highlighted differences in the capabilities of American and British naval aircraft. British Firefly I fighter-bombers could remain airborne for only around two hours, half the endurance of an American A-1 Skyraider attack plane and less than that of a F4U Corsair fighter-bomber. As a result, *Triumph* could not strike targets as deep in North Korea as could the aircraft from *Valley Forge* unless the task force moved closer to the coast, which the Americans were reluctant to do given the threat of mines and possible submarines. Admiral Hoskins thus "made clear" to *Triumph*'s captain that in future combined operations the British would be relegated to flying defensive combat air patrol and aerial surveillance patrols for the task force, while the Americans would handle offensive strikes inland. Admiral Andrewes described this decision as "galling but unquestionably correct." The two carriers continued to operate together throughout July. On 28 July, a US Air Force (USAF) B-29 bomber shot down an RN fighter, highlighting the difficulty of aerial recognition in multinational air operations. The USAF was transitioning to a new Identification Friend or Foe system, which was not compatible with British equipment. The British pilot survived the episode, and the overall USAF command, Far East Air Force, apologized for the incident but asked that British aircraft stay outside of B-29 machine-gun range in the future. Far East Air Force also provided images of British aircraft to its squadrons to aid aerial identification.[18]

By early August, North Korean forces controlled most of South Korea, forcing the remaining South Korean and American troops into defensive positions centered on Pusan. UN air operations therefore prioritized close air support for coalition ground forces and interdiction of North Korean supply routes. At sea, this interdiction effort was divided between the Korean east and west coasts. On the east coast, a rotating series of USN fleet carriers and their escorts enforced the blockade, provided close air support, and struck North Korean targets. On the west coast, British light carriers, American light and escort carriers, and one Australian light carrier performed similar functions, though typically only one carrier was on station at any given time.[19]

In September, American and South Korean troops launched an amphibious assault at Inchon well behind the front lines around Pusan. American and British carriers provided air support for the landings and spotting services for naval gunfire. British and American cruisers bombarded Inchon in the days leading up to the invasion. On the day of the landings, British Fireflies spotted naval gunfire from two British cruisers while American aircraft spotted for American ships. Combined with a counteroffensive by MacArthur's US 8th Army out of the Pusan perimeter, the Inchon landings broke the back of the North Korean invasion and the overly extended Communist supply lines. The UN forces swiftly drove the North Korean Army back across the thirty-eighth parallel. In the wake of these victories, the UN and Washington authorized General MacArthur to advance into North Korea with the goal of unifying the peninsula by force, an expansion of the original objective of securing South Korea from invasion. As MacArthur's troops drove deeper into North Korea in October, aerial interdiction targets dried up as supply routes fell into UN hands.[20]

However, Chinese intervention in late October and November dramatically expanded calls for close air support. Mao and other Chinese leaders saw Chinese participation in the Korean War as an opportunity to more firmly establish the PRC, while enhancing China's status in the region. Chinese forces in late November accordingly launched a devastating attack on advancing UN forces. The Chinese drove south, inflicting heavy casualties and taking Seoul in early January 1951. UN aircraft provided close air support for retreating UN forces, most famously for the withdrawal of the 1st Marine Division during the Battle of the Chosin Reservoir.[21]

Close air support operations required carrier-based aircraft to work with the land-based Joint Operations Center (JOC), manned by 8th Army and 5th Air Force personnel. The JOC was established in July 1950 and was responsible for receiving requests for close air support from corps and division headquarters and allocating aircraft to meet these requests. The challenge was to handle aircraft from different services (USAF, USN, US Marine Corps [USMC], RN, Royal Australian Air Force, and South African Air Force) to support the various ground formations (US Army, Republic of Korea Army, USMC, and UN contingents). Procedural and operational differences caused problems for this multiservice, multinational endeavor. For example, the JOC was designed to handle a steady stream of close air support missions. However, operations on USN fleet carriers were structured around launching large numbers of aircraft at a time (deck-load strikes), which tended to overwhelm the JOC allocation system. Furthermore, the navy and marines viewed close air support through a different framework than the air force. Whereas the air force typically viewed close air support as occurring behind the immediate front lines, the navy and marines in the Second World War had used close air support as a substitute for artillery in direct support of troops in contact. The air force had prioritized strategic bombing, not close air support, in training, funding, and doctrine in the years after the Second World War. The 5th Air Force fighter squadrons in Japan in 1950 were focused on air defense and bomber-escort missions, not on close air support. The bruising interservice battles between the navy and air force in Washington in the late 1940s meant each service approached the air war in Korea wary of the other.[22]

Doctrinal differences aside, communications also caused problems with close air support and overall air operations. Aircraft carriers preferred to operate in radio silence to avoid giving away their position, which hindered communication between the carriers at sea and the JOC. The radio equipment in British naval aircraft had fewer available channels than the equipment in their American counterparts, which made communications with the JOC, the British carriers, and a forward air controller difficult. The British were also behind the Americans in installing ultra-high-frequency, UHF, radios into their ships and planes, a lag that forced more radio traffic onto crowded very-high-frequency, VHF, channels. Furthermore, British carriers struggled at times to decode the sheer volume of American signals and messages. British warships carried fewer communications personnel than their American counterparts, which only exacerbated the problem. One RN report complained that the USN demonstrated "a complete disregard of economy in communications." Even USN reports acknowledged that the scale of message traffic was excessive at times and that "we have much to learn in communications from the British." Although the communication difficulties were manageable, they proved to be a recurring concern throughout the conflict.[23]

By late February 1951, the 8th Army had halted the Chinese offensive south of Seoul and soon began a series of counterattacks. Carrier aircraft supported these ground offensives through air strikes and spotting for naval gunfire. On the west coast, British and American carriers continued their rotation, while the large fleet carriers of Task Force 77 operated off the east coast. However, in early April the USN redeployed Task Force 77 to the waters between China and Formosa (Taiwan) due to concerns about a possible Chinese Communist assault on the Chinese Nationalists on Formosa. Two of the west coast carriers, USS *Bataan* and HMS *Theseus,* moved to the peninsula's east coast to take Task Force 77's place. For a week, the two carriers operated

Table 7.3.   Korean War Combined Carrier Operations, 1950–1951

Dates	Carriers Involved
July–August 1950	USS *Valley Forge*, HMS *Triumph*
April 1951	USS *Bataan*, HMS *Theseus*
May 1951	USS *Bataan*, HMS *Glory*
July 1951	USS *Sicily*, HMS *Glory*

Sources: Richard P. Hallion, *The Naval Air War in Korea* (Tuscaloosa: University of Alabama Press, 2011), 154–55; David Hobbs, *The British Carrier Strike Fleet after 1945* (Annapolis, MD: Naval Institute Press, 2015), 37–44; John Lansdown, *With the Carriers in Korea: The Fleet Air Arm Story, 1950–1953* (Cheshire, UK: Crecy, 1997), 110, 136, 157.

together in one of the few combined (multinational) carrier operations of the war, as shown in table 7.3.

The two carriers took turns providing combat air patrols and antisubmarine patrols for the formation and exchanged liaison officers to facilitate operations. The two ships' companies and air groups developed a healthy sense of competition, which brought about a "general speeding up of flying operations." Due to differences in range and speed, aircraft from the two ships did not operate in combined formations except when covering a downed pilot. The two carriers were screened by American, Canadian, British, and Australian escorts, a "truly United Nations" formation, according to the *Bataan*'s action report. The otherwise successful cooperation was marred on 10 April when two land-based USMC Corsairs mistakenly attacked a Sea Fury flight from *Theseus*, although no one was injured in the incident.[24]

Aside from the early days of the war when *Valley Forge* and *Triumph* operated together, combined operations involved Fleet Air Arm and USMC squadrons. For example, *Bataan* was operating Marine Fighter Squadron 312 during the April 1951 operations with *Theseus*. American practice was to keep the same marine squadron operating on the west coast for weeks, moving from carrier to carrier, rather than keeping the squadron with a single carrier. Over time, this approach limited the squadron's ability to connect with an individual carrier. Rear Admiral Eric Clifford, the senior British admiral in Korean waters in 1953, wrote that operating marine aircraft on USN ships "leads to certain difficulties from the Command point of view. Although there is no positive proof, it would appear that the Commanding Officer cannot really do as he pleases in so far as the Marine Squadron is concerned. As might be expected when dealing with another Service, there appears to be some reluctance to give orders (and enforce their compliance) on certain operational matters about which the Squadron may be somewhat 'touchy.'"[25]

By June 1951, the front lines stabilized just north of the thirty-eighth parallel. Chinese advances in April and early May were countered by UN counterattacks in late May. That June, the Far East Air Force launched Operation Strangle, an effort to interdict Communist supply lines by attacking the road network. Sectors of interdiction were separated between carrier aircraft and 5th Air Force aircraft. In August, Communist success in repairing road damage forced a shift in interdiction efforts from road to rail targets. However, the interdiction campaign was hampered by bad weather, which often precluded flying, and an extremely limited night-flying capability. According to one report, night flying accounted for only 4 percent of the air effort in Korea. As a result, Chinese and North Korean forces and supplies moved primarily at night, as did the more than 50,000 North Korean workers tasked with road and rail repair work. As large-scale ground combat became less frequent, the supply requirements of Communist forces also fell, further complicating air interdiction efforts.[26]

UN air forces continued these efforts throughout 1951 and 1952. Although air attacks did affect the Communist ability to move supplies, they appeared to have little influence on the armistice negotiations, which had begun in July 1951. In the words of one historian, "By the middle of 1952, its [the USN's] aircraft had been involved in twelve different phases of interdiction, none of which was deemed very successful."[27] UN aircraft were simply unable to stop the flow of supplies to Communist ground forces, especially given Communist manpower-intensive logistics and low supply needs. At most, aerial efforts could reduce the rate of resupply, which in turn slowed down the Communists' ability to prepare large ground offensives.

As a result, in the summer of 1952 USAF planners shifted the emphasis of the air campaign to destroying economic and industrial targets in North Korea. These strikes were intended to put pressure on the Communists to make concessions in the armistice talks. The attacks enjoyed some success. By late June, USN, Far East Air Force, and RN attacks knocked out 90 percent of North Korea's electrical power capacity. On 11 July, a large, multinational air strike hit targets in and around Pyongyang in Operation Pressure Pump. More than 1,200 UN sorties sought to overwhelm Pyongyang's heavy antiaircraft defenses, which included radar-controlled antiaircraft guns. The 5th Air

Force coordinated this operation involving aircraft from the USAF, USMC, USN, RN, and Royal Australian Air Force.[28]

In contrast to the overall focus on interdiction and destruction strikes in late 1951 and 1952, the British, American, and Australian light and escort carriers on the west coast continued to give significant attention to close air support. For example, on a typical operating day in January 1952 the USMC fighter squadron on USS *Badoeng Strait* flew forty-one sorties, including twenty-three offensive sorties, eight armed reconnaissance sorties, and ten combat air patrols over the carrier.[29]

In addition to performing air strikes, carrier aircraft regularly spotted gunfire from warships off the Korean coast. In 1950, the British and American navies used different procedures for spotting naval gunfire, but the Korean War forced RN units to adopt American procedures. British observers generally felt that American aerial spotting was "uneven" due to less training and practice in spotting procedures. On several occasions, American ships and pilots were unfamiliar with their own spotting procedure, and British personnel consequently provided lectures and training to remedy the situation. The USN was more comfortable with firing on unobserved targets, resulting, from the British perspective, in less emphasis on spotting. One British report noted that the marine squadron (VMF-323) that joined USS *Sicily* in June 1951 "had little or no previous training in bombardment." In contrast, British carrier pilots regularly practiced spotting naval gunfire during their predeployment training. On the whole, Commonwealth ships and some American ships preferred the spotting done from British aircraft due to the RN's emphasis on this activity.[30]

Multinational spotting operations were also complicated by differences in accents. A British frigate picked up a USMC pilot shot down in October 1951 but found that it was having difficulty communicating with the marine aircraft providing spotting services. The pilot, Captain G. C. Armstrong, reported that "most of the trouble came from voice accent," and he took over as air controller by directing several spotting missions, "all with good results." A US Pacific Fleet report complained that American operators on spotting radio circuits "had difficulty understanding voice transmissions from British fire-support ships."[31] Accents were a problem not just in combined operations. A history of RN signaling notes that "British radio operators had problems understanding accents from the different parts of the British Isles rather than those of the Americans."[32]

Although both British and American sources comment on language difficulties, only British sources regularly mention differences in personnel policy. British reports specifically criticized the frequent turnover of American pilots, which in the British view reduced squadron performance. The marine squadrons embarked on American carriers on the Korean west coast relied heavily on reservist pilots, who were rotated in and out of the squadron at regular intervals. This approach meant that the squadron always had the benefit of a crop of veteran pilots but also that there was a constant influx of newer pilots with less experience in Korea. In the words of one British report at the time, this personnel policy "results in the squadron never being worked up."[33]

The Americans also rotated their senior leaders more frequently than did the British. Eight different USN admirals commanded the blockade and escort force (Task Force 95) during the three years of the Korean War. The east coast fast-carrier command (Task Force 77) was held by thirteen different American admirals during the conflict. In contrast, only three British admirals served as flag officer, second in command, Far East Station, the senior British command in Korea. On the whole, it seems the USN wanted to give combat experience to a large number of its admirals. This approach was beneficial for the USN in the long term, and these frequent rotations meant that commanders who caused problems for multinational operations, such as Rear Admiral George Dyer, did not remain in command too long. However, the regular turnover in naval commands in Korea made the British feel as though they had to constantly explain local conditions and procedures to newly arriving American admirals.[34]

Working as part of the same task force on the Korean west coast led to a healthy sense of competition between carriers, especially for the RN. British carriers regularly worked to outdo one another and their American counterparts in the number of sorties flown in a single day. In the summer of 1951, British light fleet carriers (the type used in Korea) flew roughly 50 sorties a day under good weather for flying, comparable to the number of sorties flown by the USS *Bataan,* a US light fleet carrier, earlier that year. *Theseus* held the record in 1951 with 60 sorties a day, until *Glory* achieved 105 in March 1952, which *Ocean* topped with 123 in May 1952. Not to be outdone, *Glory* matched *Ocean* in April 1953. These high sortie rates were achieved on good-weather days by launching planes in the predawn hours and having each pilot fly four sorties throughout the day. This level of effort was not sustainable but did reflect a strong competitive spirit. When *Bataan* operated with *Theseus* and then *Glory* in April and May 1951, personnel on all three ships worked to speed up launching and landing operations, again reflecting a sense of friendly rivalry.[35]

The RN's high sortie rates were even more impressive given that the Americans carried more pilots per aircraft than the British. American fleet carriers had roughly three pilots for every two aircraft, and American escort carriers had an almost two-to-one ratio of pilots to planes. In contrast, on British and Australian carriers the ratio was closer to one to one. The British achieved their record-breaking numbers through a maximum effort but recognized that the American approach was better suited for sustained, intensive operations. The differences in pilot-to-plane ratios reflected technical differences in warship design. American carriers enjoyed comparatively larger personnel quarters than their British counterparts. The USN in the interwar period planned to operate carriers in the western Pacific against Japan far from fleet bases, which in turn meant carriers needed spare pilots to continue fighting amid the attrition of combat. In contrast, British light fleet carriers were designed during the Second World War as austere ships in the hopes of reducing construction times to meet wartime demands for carriers. As a result, they were cramped and lacked the space to carry large reserves of pilots. Such technical differences highlighted how each navy's unique needs shaped warship design and operational capabilities. Both approaches made sense in their appropriate context.[36]

Conclusion

Multinational air efforts invariably draw out differences between services. Combined carrier operations in the Korean War revealed differences in spotting procedure, language, personnel policy, and warship design. Even in services with as much in common as the RN, the Royal Australian Navy, and the USN, fighting together put a spotlight on areas of divergence. However, the navies generally overcame these potential rifts and found workable solutions, largely through relations between midlevel and junior officers.

The USN geographically separated itself from British and Australian carriers by placing them on the west coast of Korea in the Yellow Sea, while the large American fleet carriers stayed off the east coast in the Sea of Japan. This approach mirrored the general approach of the Second World War, where each nation assumed responsibility for specific geographic regions. For example, amphibious assaults were divided into American and British sectors in North Africa, the Mediterranean, and Normandy. A similar policy applied on the ground in Korea as the 8th Army gave the 1st Commonwealth Division (formed in July 1951) a separate part of the front line. The consequence of putting the Commonwealth carriers on the west coast was that only a small portion of the USN forces operating in Korea regularly worked with the British and Commonwealth. Specifically, American escort and light carriers gained experience in multinational operations that the larger fleet carriers did not. In contrast, all the British and Commonwealth naval units received a strong dose of combined operations.

Multinational carrier operations certainly contributed to the outcome of the Korean War, though their international flavor was not decisive on its own. The presence of British and Australian carriers in Korea made the USN's task easier, though the Americans still flew the preponderance of carrier-launched sorties over the course of the war. The Korean War did have an impact on long-term relations between the navies involved. The conflict highlighted the value of regular combined exercises as a tool for retaining interoperability. The experience of operating in multinational formations convinced many British and American officers that such operations would be the pattern in the future. The war also highlighted the value of shared tactical publications, such as the NATO common operating procedures, which were then in development.[37]

Historical and organizational differences inevitably created potential roadblocks to effective combined carrier operations in Korea. When issues such as differences in spotting procedure arose, naval personnel typically worked out solutions at the local or theater level. Rather than raising issues at the chiefs-of-service level, commanders met and talked through their problems. Regular exchanges of liaison personnel and operational notes proved invaluable for west coast carrier operations, as did continuity in personnel. Whereas American admirals rotated frequently, British admirals did not. The bulk of west coast carrier tours of duty were handled by two British and four American carriers. By 1953, one British report was referring to the American west coast carriers as "our old friends." This degree of continuity and a preference for local solutions to local problems helped improve the efficiency of west coast carrier operations.[38]

The war also helped solidify the foundations for ongoing cooperation between the American, British, Canadian, and Australian navies during the

Cold War. In the field of naval aviation, growing use of American naval aircraft helped improve interoperability. In the 1950s, the Canadian navy bought American Banshee fighters and Tracker antisubmarine aircraft. In the 1960s, the Australians bought Trackers as well as Skyhawk attack planes, while the RN acquired American Phantom fighters. USN ships and aircraft regularly exercised with their British and Commonwealth counterparts in NATO and Southeast Asian Treaty Organization exercises. Formal personnel exchanges, for both operational and educational duty, helped maintain close relations between these four navies. Without such links and regular contact, interoperability tended to gradually fade away. Like any skill, combined carrier operations required regular practice.

NOTES

1. The views expressed in this chapter are those of the author and do not reflect the official policy or position of the US government, the Department of Defense, or Air University. The author thanks Frank Blazich for his comments on an early draft of this chapter.

2. Samuel Eliot Morison, *Victory in the Pacific, 1945*, vol. 14 of *History of United States Naval Operations in World War II* (Edison, NJ: Castle Books, 2001), 362–67.

3. This chapter references three different types of aircraft carriers. Fleet carriers (CVs) were the largest carriers, designed to operate large air groups in offensive and defensive roles and sufficiently fast to keep up with a fleet. Light fleet carriers (CVLs) were smaller than fleet carriers and carried smaller air groups but were also built to accompany the fleet. Escort carriers (CVEs) were slower and smaller, designed for supporting roles such as protecting convoys and providing air support for amphibious operations.

4. The term *combined* in this chapter refers to multinational, whereas the term *joint* refers to multiservice from a single nation, in keeping with US military parlance at the time.

5. The value of trust in multinational operations is thoroughly examined in Gary Weir and Sandra Doyle, eds., *You Cannot Surge Trust: Combined Naval Operations of the Royal Australian Navy, Canadian Navy, Royal Navy, and United States Navy, 1991–2003* (Washington, DC: Naval History and Heritage Command, 2013).

6. Peter Clarke, *The Last Thousand Days of the British Empire: Churchill, Roosevelt, and the Birth of Pax Americana* (New York: Bloomsbury Press, 2008), 402; Michael Simpson, "Admiral Viscount Cunningham of Hyndhope," in *The First Sea Lords: From Fisher to Mountbatten*, ed. Malcolm Murfett (Westport, CT: Praeger, 1995), 218; UK Central Statistical Office, *Annual Abstract of Statistics No. 90* (London: Her Majesty's Stationery Office, 1953), 95. Throughout 1946 and 1947, the Attlee government considered a variety of proposals regarding the RN's appropriate postwar strength, though the general trend in size was clearly downward. One American intelligence report commented wryly in late 1947 that "because of the day to day uncertainties of recent months, forecasting the future

strength of the Royal Navy has become a hazardous enterprise for even the First Sea Lord" (Office of Naval Intelligence, "Serial 47-47, International Developments of Naval Interest," 21 November 1947, folder A8 "Intelligence Jacket #4 [9–12/47] [1 of 4]," box 162, entry P 111, Commander in Chief, US Atlantic Fleet [CINCLANT], Secret and Top Secret Correspondence, 1941–1949, Record Group [RG] 313, Records of Naval Operating Forces, National Archives [NARA] at College Park, College Park, MD). For full details of these various proposals, see Eric Grove, *From Vanguard to Trident: British Naval Policy since World War II* (Annapolis, MD: Naval Institute Press, 1987).

7. Grove, *From Vanguard to Trident,* 17; David Hobbs, *The British Carrier Strike Fleet after 1945* (Annapolis, MD: Naval Institute Press, 2015), 9.

8. Tim Benbow, "British Naval Aviation and the 'Radical Review,' 1953–55," in *British Naval Aviation: The First 100 Years,* ed. Tim Benbow (Burlington, VT: Ashgate, 2011), 126. All together, these innovations made the operation of high-performance jet aircraft from aircraft carriers possible. See Antony Preston, "A Great Navy in Decline: The Royal Navy since 1947," in *New Aspects of Naval History: Selected Papers from the 5th Naval History Symposium* (Baltimore: Nautical & Aviation Publishing, 1985), 208; Thomas C. Hone, Mark Mandeles, and Norman Friedman, *American & British Aircraft Carrier Development, 1919–1941* (Annapolis, MD: Naval Institute Press, 1999).

9. Robert H. Ferrell, *Harry S. Truman: A Life* (Columbia: University of Missouri Press, 1994), 338.

10. Herman S. Wilk, *Toward Independence: The Emergence of the U.S. Air Force, 1945–1947* (Washington, DC: Air Force History and Museums Program, 1996), 16–17.

11. William D. Leahy, diary entry for 21 May 1946, frame 339, reel 4, folder 1946, box 7, William D. Leahy Papers, Naval Historical Collection, Manuscript Division, Library of Congress (LoC), Washington, DC.

12. Jeffrey Barlow, *Revolt of the Admirals: The Fight for Naval Aviation, 1945–1950* (Washington, DC: Naval Historical Center, 1994); Allan Millett, Peter Maslowski, and William B. Feis, *For the Common Defense: A Military History of the United States from 1607 to 2012* (New York: Free Press, 2012), 450.

13. Allan Millett, *The War for Korea, 1945–1950: A House Burning* (Lawrence: University Press of Kansas, 2005); William Stueck, *The Korean War: An International History* (Princeton, NJ: Princeton University Press, 1997), 24–43.

14. Stueck, *The Korean War,* 41–46.

15. Rear Admiral Sir William Andrewes, RN, Flag Officer Second in Command, Far East (FO2FE) to Admiral Patrick Brind, RN, Commander in Chief, Far East (CinCFE), "FO2FE/2960/11, Korean War—Second Report of Proceedings, 1st to 5th July, 1950," 10 July 1950, Naval Ops in Korea I–IV, DS/Misc/12, microfilm reel 1, Papers of Admiral Sir William Andrews, Documents and Sound Section, Imperial War Museum (IWM), London.

16. Rear Admiral Walter F. Boone, USN, Commander Carrier Division Five, to Vice Admiral Russell S. Berkey, USN, Commander Seventh Fleet, "Report of Combined Exercises with British Far East Fleet during the Period 28 February–19 March 1950," 21 March 1950, folder "Commander in Chief, US Pacific Fleet (CINCPACFLT)," serial 0332, 30 April 1950, box 22, Post January 1946 Reports file, Archives Branch, Naval History and Heritage Command (NHHC), Washington, DC; "Professional Notes: British Carrier Operations,"

U.S. Naval Institute Proceedings 75, no. 3 (March 1949): 375; "Reminiscences of Vice Admiral Charles A. Pownall," USN, interview by Etta Belle Kitchen, 1989, 115–16, US Naval Institute Oral Histories, Navy Department Library, NHHC; Ray Sturtivant, *British Naval Aviation: The Fleet Air Arm, 1917–1980* (Annapolis, MD: Naval Institute Press, 1990), 139; Rear Admiral William Andrewes, RN, F02FE, to Admiral Patrick Brind, RN, CinCFE Station, "F02FE/2263/22, Combined Exercises with United States Navy—March, 1950," 12 March 1950, ADM 1/21868, The National Archives of the United Kingdom (TNA); Tactical and Staff Duties Division, "CB 3016 (50), Progress in Tactics, 1950," 26 June 1950, 144, box 12, entry UD-09D 20, British Records & Publications, 1914–1955, RG 38, Records of the Office of the Chief of Naval Operations, NARA.

17. Norman Polmar, *1946–2006,* vol. 2 of *Aircraft Carriers: A History of Carrier Aviation and Its Influence on World Events* (Washington, DC: Potomac Books, 2008), 57; Warren E. Thompson, *Naval Aviation in the Korean War* (Barnsley: Pen & Sword Aviation, 2012), 16.

18. Captain Arthur D. Torlesse, RN, Commanding Officer (CO) HMS *Triumph*, to Rear Admiral William Andrewes, RN, F02FE, "No. 02180, Report on Operations, 2nd–5th July, 1950," 21 July 1950, 2, ADM 116/6224, TNA; Captain Arthur D. Torlesse, RN, CO HMS *Triumph*, to Rear Admiral William Andrewes, RN, F02FE, "No. 02186, Report of Proceedings," 22 July 1950, ADM 116/6224, TNA; Rear Admiral William Andrewes, RN, F02FE, to Vice Admiral Patrick Brind, RN, CinCFE, "F02FE/2960/72," 16 August 1950, ADM 116/6224, TNA; Captain Lester Kimmie Rice, USN, CO USS *Valley Forge*, to Commander Seventh Fleet, "Serial 082, Report of Operations 16 July to 31 July 1950," 1950, Aviation History Branch, NHHC, at https://web.archive.org/web/20170507075722 /https://www.history.navy.mil/content/dam/nhhc/research/archives/action-reports /Korean%20War%20-%20Carrier%20Combat/Valley%20Forge%20(CV%2045) %20Action%20Reports%20for%20Korea/PDF's/cv45-j50.pdf; Hobbs, *The British Carrier Strike Fleet after 1945,* 37; Conrad Crane, *American Airpower Strategy in Korea, 1950–1953* (Lawrence: University Press of Kansas, 2000), 30; Steven Paget, *The Dynamics of Coalition Naval Warfare: The Special Relationship at Sea* (New York: Routledge, 2017), 70.

19. For a survey of the war's first year, see Allan Millett, *The War for Korea, 1950–1951: They Came from the North* (Lawrence: University Press of Kansas, 2010).

20. Stueck, *The Korean War,* 94; Crane, *American Airpower Strategy in Korea,* 44; Richard P. Hallion, *The Naval Air War in Korea* (Tuscaloosa: University of Alabama Press, 2011), 61; Paget, *The Dynamics of Coalition Naval Warfare,* 24–25.

21. Millett, *The War for Korea, 1950–1951,* 350; Lieutenant Commander A. S. Pomeroy, RN, British Naval Liaison Officer with Commander, US Naval Forces (COMNAVFE), to Air Vice Marshal Cecil A. Bouchier, RAF, Senior British Liaison Officer, Tokyo, "CSS 14, Routine Report No. 14," 11 December 1950, 114–15, ADM 116/5777, TNA (Australian Joint Copying Project reel PRO 5945).

22. Paget, *The Dynamics of Coalition Naval Warfare,* 40–41; Rear Admiral Alan K. Scott-Moncrieff, RN, F02FE, "Report of Experience in Korean Operations, January–June 1951," 27 July 1951, part III, sec. 1, ADM 116/6230, TNA; Wing Commander James E. Johnson, RAF, "Report from Korea" (1951), 4, folder "Slessor, John," box 108, Prepresidential, 1916–1952, principal file, Dwight David Eisenhower Papers, Dwight David Eisenhower Presidential Library, Abilene, KS; Hallion, *The Naval Air War in Korea,* 42–43.

23. William T. Y'Blood, *Down in the Weeds: Close Air Support in Korea* (Washington, DC: Air Force History and Museums Program, 2002), 16; Johnson, "Report from Korea," 5; Anthony Farrar-Hockley, *An Honorable Discharge*, vol. 2 of *The British Part in the Korean War* (London: Her Majesty's Stationery Office, 1995), 301; Military History Section, United Nations Command, Far East Command Headquarters, "Problems in Utilization of United Nations Forces, 8–5.1A AC," in *U.S. Army Center of Military History Historical Manuscripts Collection: The Korean War* (Wilmington, DE: Scholarly Resources, 2000), reel 12, 54–55; David Hobbs, "Inter-Allied Communication during the Korean War," in *Naval Networks: The Dominance of Communications in Maritime Operations,* ed. David Stevens (Canberra: Sea Power Centre–Australia, 2012), 181–82; CINCPACFLT, "Chapter 5 Surface Operations, Interim Evaluation Report No. 4," 1952, 5–52, folder "Chapter 5, Surface Operations No. 4, January 1952–June 1952," box 7, entry P3, Korean War Interim Evaluation Reports, 1950–53 (Interim Reports), RG 38, NARA; Scott-Moncrieff, "Report of Experience in Korean Operations, January–June 1951," 1; CINCPACFLT, "Chapter 5 Surface Operations, Interim Evaluation Report No. 6" (1953), 5–113, folder "Chapter 5 Surface Operations No. 6, February 1953–July 1953," box 8, Interim Reports, RG 38, NARA.

24. John Lansdown, *With the Carriers in Korea: The Fleet Air Arm Story, 1950–1953* (Cheshire, UK: Crecy, 1997), 110; Captain Arthur S. Bolt, RN, CO HMS *Theseus,* to Rear Admiral Alan K. Scott-Moncrieff, RN, F02FE, "No. 191/1," 19 April 1951, 1, ADM 116/6224, TNA; Major Frank H. Presley, USMC, CO Marine Fighter Squadron 312, to Commandant of the Marine Corps (CMC), "Serial 0380, Historical Diary for 1 April to 30 April 1951," 3 May 1951, 10–13, Korean War Project, at https://www.koreanwar2.0rg/kwp2/usmc/032 /m032_cd10_1951_04_857.pdf; Captain William Miller, USN, CO USS *Bataan,* to Chief of Naval Operations (CNO), "Serial 050, Action Report; Period 8 April 1951–11 May 1951, CVL29, A16–13," 12 June, 1951, I–1, Aviation History Branch, NHHC, at https://web. archive.org/web/20121025141932/http://www.history.navy.mil/a-korea/cv129-am51.pdf.

25. Rear Admiral Eric G. A. Clifford, RN, F02FE, "Report of Experience in Korean Operations, July 1952–April 1953," 15 July 1953, part III, sec. 1, file 1926-102/11 Pt. 2, "Reports of Proceedings—Commander Canadian Destroyers—Far East, 1951–1953," vol. 8204, RG 24-D-1-c, Library and Archives Canada, Ottawa.

26. Ibid., 82–84; Johnson, "Report from Korea," 4; Paget, *The Dynamics of Coalition Naval Warfare,* 32.

27. Crane, *American Airpower Strategy in Korea,* 113.

28. Ibid., 116–19; Hobbs, *The British Carrier Strike Fleet after 1945,* 54–55; Scott-Moncrieff, "Report of Experience in Korean Operations, January–June 1951," part III, sec. 1; Lansdown, *With the Carriers in Korea,* 262.

29. Hallion, *The Naval Air War in Korea,* 92; Captain Rudolph Lee "Roy" Johnson, USN, CO USS *Badoeng Strait,* to CNO, "Action Report 7 January–16 January 1952 and 25 January–6 February 1952, A12/OF30," 17 March 1952, 3–4, Aviation History Branch, NHHC, at https://web.archive.org/web/20070710134136/http://history.navy.mil:80/a -korea/cve116-17-28dec51.pdf.

30. Paget, *The Dynamics of Coalition Naval Warfare,* 55–64; Scott-Moncrieff, "Report of Experience in Korean Operations, January–June 1951," part III, sec. 3; Australian Naval Aviation Museum, *Flying Stations: A Story of Australian Naval Aviation* (Sydney: Allen & Unwin, 1998), 88; Alan Leahy, interview by Richard McDonough, 28 September 2010,

reel 9, catalogue 32843, Sound Archive, IWM; Alexander Murray Scrimgeour, interview by Toby Books, March 2008, catalogue 31041, Sound Archive, IWM; Hobbs, *The British Carrier Strike Fleet after 1945,* 65–66.

31. CINCPACFLT, "Interim Evaluation Report No. 1, Project No. II.B, Communications," 1950, 1442, folder "Report No. 1, June 1950–November 1950," vol. VIII, box 2, Interim Reports, RG 38, NARA.

32. Enclosure to Lieutenant Colonel Manual Brilliant, USMC, CO Marine Fighter Squadron 212, to CMC, "Serial 002-51, Historical Diary for 1–31 October 1951," 12 November 1951, 55, appendix V, Korean War Project, at https://www.koreanwar2.0rg/kwp2/usmc/027/m027_cd08_1951_10_743.pdf; Barrie Kent, *Signal! A History of Signalling in the Royal Navy* (Clanfield, UK: Hyden House, 1993), 172.

33. Paget, *The Dynamics of Coalition Naval Warfare,* 61–62; Clifford, "Report of Experience in Korean Operations, July 1952–April 1953," part III, sec. 1.

34. Hallion, *The Naval Air War in Korea,* 138; Gordon Rottman, *Korean War Order of Battle: United States, United Nations, and Communist Ground, Naval, and Air Forces, 1950–1953* (Westport, CT: Praeger, 2002), 93; Corbin Williamson, "Fighting with Friends: Coalition Warfare in Korean Waters, 1950–1953," *Joint Force Quarterly* 83 (October 2016): 99–104.

35. Hobbs, *The British Carrier Strike Fleet after 1945,* 44, 47, 51–53, 59; CINCPACFLT, "Chapter 3 Carrier Operations, Interim Evaluation Report No. 4," 1952, 3–30, folder "Chapter 3, Carrier Operations No. 4 January 1952–June 1952," box 4, Interim Reports, RG 38, NARA; Major Donald P. Frame, USMC, CO Marine Fighter Squadron 312, to CMC, "Serial 0373, Historical Diary for 1 March to 31 March 1951," 3 April 1951, 7, Korean War Project, at https://www.koreanwar2.0rg/kwp2/usmc/032/m032_cd10_1951_03_856.pdf.

36. Vice Admiral William Andrewes, RN, F02FE, "Report of Experience in Korean Operations, July–December 1950," March 31, 1951, part III, sec. 1, ADM 116/6230, TNA; Lieutenant Colonel Joseph A. Gray, USMC, CO Marine Fighter Squadron 212, to CMC, "Serial 001-52, Historical Diary for 1–31 December 1951," 27 February 1952, 35, 42–43, appendixes 1 and 3, Korean War Project, at https://www.koreanwar2.0rg/kwp2/usmc/027/m027_cd08_1951_12_745.pdf; Russ Mallace, "HMS Ocean in the Korean War," n.d., Royal British Legion, at http://branches.britishlegion.org.uk/branches/byfield/stories/hms-ocean-in-the-korean-war; Norman Friedman, *British Carrier Aviation: The Evolution of the Ships and Their Aircraft* (Annapolis, MD: Naval Institute Press, 1998), 218–29; Norman Friedman, *U.S. Aircraft Carriers: An Illustrated Design History* (Annapolis, MD: Naval Institute Press, 1983), 13, 65, 121; David Hobbs, *British Aircraft Carriers: Design, Development, and Service Histories* (Barnsley, UK: Seaforth, 2013), 183.

37. Corbin Williamson, "We Are Still One Fleet: U.S. Navy Relations with the British, Canadian, and Australian Navies, 1945–1953," PhD diss., Ohio State University, 2015, 276–77.

38. Clifford, "Report of Experience in Korean Operations, July 1952–April 1953," part III, sec. 1.

Magpies and Eagles

Number 2 Squadron, Royal Australian Air Force, and the Experience of Coalition Warfare in Vietnam

Steven Paget

Much has been written about the Vietnam War, but as Jonathan Colman and J. J. Widen have outlined, "many texts" overlook the fact that it was an "example of coalition warfare."[1] Indeed, despite the wealth of literature on the use of air power during the conflict, there has been a relative dearth of studies on the Australian contribution.[2] By comparison to the American commitment, the Royal Australian Air Force (RAAF) made a modest contribution to the air power effort in Vietnam. The RAAF deployed 2 Squadron's Canberra bombers, 9 Squadron's Iroquois helicopters, and 35 Squadron's Caribou transport aircraft to Vietnam. Australian Air Defence Guards and the 5th Airfield Construction Squadron were also deployed, and extensive sorties were conducted by the transport fleet to support operations.[3] In addition, forty RAAF officers served as forward air controllers or F-4 Phantom pilots on exchange with the Americans, and a number of 9 Squadron pilots flew with the 135th Assault Helicopter Company to aid the transition to UH-1Hs.[4] Notably, 2 Squadron—employing the call sign "Magpie"—conducted operations in South Vietnam between April 1967 and June 1971 and, according to statistics compiled by the United States, managed to achieve the best bomb-damage-assessment performance of any allied squadron.[5]

Irrespective of the significance of the Australian contribution to the out-come of the conflict, the increasing prevalence of multinational operations throughout the twentieth century and into the twenty-first century has ensured that the coalition nature of the Vietnam War is noteworthy. In par-ticular, Australia and the United States have a long-standing history of mul-tinational operations dating back to the First World War. The close and enduring relationship was encapsulated in a review published in 2002 of strategic-level interoperability between Australia and the United States: "For Australia and the United States, the ANZUS treaty [Australia, New Zealand, United States Security Treaty of 1951] has always symbolized and formalized a close alignment of enduring shared values and common interests. The extent of commitment to shared values and interest (peace, democracy, regional stability) were [sic] underscored by the alongside-service of Austra-lian and U.S. military forces in both World Wars, Korea, Vietnam, the Persian Gulf War, Somalia, East Timor, and, most recently, Afghanistan and Iraq."[6]

As multinational operations have become "part of the 'core' business" of the Australian Defence Force, historical experience is of great significance.[7] The cultural similarities between Australians and Americans—typified dur-ing the Second World War by the Australian prime minister John Curtin's description of US service personnel as "visitors who speak like us, think like us and fight like us"—has limited discussion of the challenges of enacting interoperability.[8]

Using the experience of 2 Squadron as a lens, this chapter addresses the dynamics of conducting multinational air operations over South Vietnam. It considers the practical and conceptual challenges posed by 2 Squadron's inte-gration with the US Air Force (USAF) and the conduct of operations in the multinational environment. It also analyzes key issues such as command and control, logistics, rules of engagement, standardization, and tactics, tech-niques and procedures to provide insights into multinational operations in the air environment.

The experiences of 2 Squadron serve as an exemplar of the benefits and challenges presented by the conduct of multinational air operations during the Vietnam War. The conflict demonstrated that clear command-and-control arrangements and well-defined rules of engagement (ROE) were required to build a platform for cohesion. However, the challenges of integration and vari-ations in capabilities and ROE meant that goodwill and mutual understanding were also necessary. The RAAF's involvement in the Vietnam War was particularly noteworthy in that the air force, unlike the Royal Australian Navy (RAN), did not have the benefit of operating standardized platforms in theater, as in the case of 2 Squadron's Canberra bombers. The squadron's operations demonstrated that although there are obvious benefits to standardization, there are advantages—albeit sometimes unexpected—from utilizing a variety of capabilities. Nevertheless, the operation of nonstandardized platforms by 2 Squadron necessitated an acclimatization period for its USAF counterparts as they adjusted to the Canberra's capabilities and idiosyncrasies in operating styles. Although the two air forces were able to develop an effective equilibrium and operated successfully together in Vietnam, a number of obstacles had to be overcome to facilitate cohesion.

COMMAND AND CONTROL: SIMPLICITY IS KEY

Despina Tramoundanis has written that Australian forces "fought as adjuncts of United States force elements and not as a unified force" during the Vietnam War.[9] From the Australian perspective, 2 Squadron was considered an "independent, self-accounting unit" for operations in Vietnam.[10] Arrangements for 2 Squadron were relatively straightforward in that operational command was retained by the RAAF, and operational control was passed to the US 7th Air Force.[11] When 2 Squadron arrived at Phan Rang, where it was to be based, Air Commodore Jack Dowling, commander of the RAAF component in Vietnam, formally handed over operational control to General William Momyer, commander, 7th Air Force.[12] Administrative control of the squadron was exercised by the commander of Australian Force Vietnam, who was required to inform the Australian Department of Air regarding RAAF operations on a monthly basis.[13]

The nature of the command arrangements meant that "it was little beyond unit command that mattered" in practical terms for the Australians.[14] The squadron was integrated into the existing US Tactical Air Control System for operations, with fragmentary orders being issued by the Tactical Air Control Center.[15] Despite receiving operational tasking from 7th Air Force headquarters, the Australians were subordinate to but not directly commanded by the commanding officer of the 35th Tactical Fighter Wing (TFW) because 2 Squadron was situated alongside a number of other squadrons at Phan Rang.[16]

Alan Stephens has noted: "As a very small contributor to massive strike and airlift campaigns, the RAAF had limited influence over tasking, and none over policy. Nevertheless, the excellent reputations quickly earned by both squadrons [2 and 35] at least meant that when RAAF commanders did raise concerns about roles and missions their opinions were treated with respect."[17] Colonel Frank Gailer Jr., who served as commanding officer of the 35th TFW between September 1968 and August 1969, confirmed that the Americans and Australians alike were "very receptive" to the opinions of their respective counterparts.[18] As a testament to the closeness of the relationship during Wing Commander David Evans's time as 2 Squadron's commanding officer, he was seated alongside the 35th TFW's commanding officer at the daily briefings rather than with the other squadron commanders due to his designated status as the senior national Australian officer.[19]

Past experience of operating with the United States—such as during the Korean War—undoubtedly eased the process of integration, but the RAAF still had to demonstrate credibility during the initial commitment to Vietnam.[20] As 2 Squadron slotted into the existing American command structure and "got on with the job with only minimal interference or direct oversight and direction from external bodies," the arrangements essentially caused little difficulty.[21] One RAAF pilot pointed out: "Were it not for basic administration, uniforms and aircraft, 2 Sq[uadro]n operations were simply an integrated part of the larger USAF and US military."[22]

NOT ALL CANBERRAS WERE CREATED EQUAL: THE IMPORTANCE OF STANDARDIZATION

Although clear command-and-control arrangements set a platform for cooperation, the integration of 2 Squadron's Canberra aircraft created a number of challenges. In the wake of the Second World War, it was evident that interoperability would be an important factor in future military operations. The creation of the first Research, Development, and Standardization Group of US Army Materiel Command emphasized that defense cooperation, in particular standardization, would be an important driver for defense procurement.[23] Although the Americans initially focused on standardization with Canada and the United Kingdom, National Security Council Report 6109 avowed on 16 January 1961, "The United States should continue to urge Australia progressively to standardize its military equipment using United States standards and continue to facilitate, as appropriate, the purchase by Australia of United States equipment for its own forces."[24]

That sentiment was replicated in Australia. Sir Percy C. Spender, Australian ambassador to the United States, visited the Pentagon in 1957 and declared: "The inevitable conclusion is that our destinies are intertwined—if war does break out, our forces will be engaged together. To wait until such a time would be foolhardy and tragic."[25] The alignment of thinking on standardization was borne out in Australia's procurement decisions. The most notable decision, from the RAAF's perspective, was the purchase of the F-111 aircraft, which was announced in 1963.[26] Ian Bellamy and James Richardson contended that "for the first time [Australia] committed itself to purchase an aircraft in the forefront of advanced technology" with the decision to purchase the F-111s.[27] More significantly, the F-111s that Australia purchased were a variation of an American-designed and built aircraft, which represented a major milestone in standardization with the United States. Because of this purchase, combined with the decision to acquire Charles F. Adams class guided-missile destroyers for the RAN, Edward Clark, the American ambassador to Australia, affirmed in 1966 that Australia had "been most cooperative with us in the field of procurement, against rather strong competition and political pressure from the British."[28]

Unlike the RAN, which, with the exception of one vessel, was able to deploy standardized ships to operate alongside the US Navy, the RAAF was unable to put their new acquisitions to the test in Vietnam as various delays meant that the purchased F-111s did not come into service with the Australians until 1973.[29] Instead, it was the British-designed Canberra bombers of 2 Squadron that would see service alongside the Americans in Vietnam.[30] A memorandum from the US Defense Attaché Office suggested that the older aircraft, rather than being detrimental, might be preferential for wartime operations, observing: "Deployment of Canberra Sqdn is ideal for RAAF as [these] aircraft are considered obsolescent and will be replaced by F-111. RAAF will run out of crews long before they run out of Canberra aircraft."[31]

US Military Assistance Command, Vietnam (MACV), did not anticipate the British-designed Canberras to be problematic because it was believed that they would be compatible with the Martin B-57, which was an American variation of the same aircraft type. In reality, however, the B-57s were "effectively an entirely separate type."[32] Colonel W. C. Platt, director of training of Pacific Air Forces, outlined the difference: "Essentially, the B-57 performs like a fighter-bomber in an air-to-ground role and is capable of delivering a wide variety of munitions utilizing the several fighter-bomber tactics. On the other hand, the Canberra carries considerably fewer types of munitions and is essentially a straight and level bombing platform."[33] Not only did the Canberra differ from the USAF's B-57, but, as the commander of Australian Force Vietnam noted in 1970, 2 Squadron was "unique to the theatre in possessing the only low and medium level bombing capacity."[34]

Number 2 Squadron commenced operational flying on 23 April 1967 and was initially confined to Skyspot bombing missions controlled by a ground-based radar-guided bombing system that guided aircraft to their targets. The squadron had been introduced to Skyspot bombing procedures during briefings and familiarization flights for each crew during the preceding week.[35] It proved competent in that role, but Wing Commander Rolf Aronsen pushed for his crews to be allowed to conduct daylight bombing controlled by a forward air controller (FAC).[36] A postconflict Project CHECO report acknowledged: "Not only did this policy [limiting 2 Squadron to Skyspot missions] prevent the pilots from realizing their potential, but it also stifled morale, because the MSQ [Combat Skyspot] night missions gave them little BDA [bomb-damage-assessment] feedback."[37] Importantly, the change in operations was in part facilitated by the efforts of Wing Commander Tony Powell of the RAAF. With 700 hours of flying time as a FAC in the USAF as well as the experience of serving as the deputy director of the USAF Direct Air Support Center Alpha and the commander of the USAF Tactical Air Control Party attached to Australian Task Force headquarters, Powell was able to represent convincingly the merits of the Canberra. As a result, reciprocal visits were arranged between 2 Squadron personnel and Direct Air Support Centers and Tactical Air Control Parties.[38]

Number 2 Squadron eventually completed its first FAC-controlled bombing mission against a suspected assembly area in II Corps on 25 June 1967, a mission that was reported to have vindicated that the Canberra was "suitable for this role."[39] Importantly, FAC procedures were produced and disseminated to 2 Squadron by the 7th Air Force.[40] The nature of 2 Squadron operations continued to change as the 35th TFW became imbued with confidence about the Canberra's utility in the daylight visual-bombing role. When Wing Commander Evans took command in 1967, 2 Squadron was flying six Skyspot missions and two visual-bombing sorties per day. That ratio was reversed by the time he relinquished command in 1968.[41] As Australian involvement continued, progressively fewer Skyspot sorties were flown as low-level, visual bombing was favored.[42] When weather conditions or other factors led to the cancellation of visual bombing, however, Skyspot operations were conducted instead. Despite initial USAF reservations about the Canberra's usefulness in the daylight visual-bombing role, given the potential vulnerabilities of the aircraft when conducting low-level strikes, the aircraft proved its worth in the relatively low-threat environment over South Vietnam in which small-arms fire and medium antiaircraft guns were the predominant risks.[43]

KNOW THY FRIEND: RAAF TACTICS, TECHNIQUES, AND PROCEDURES

As a level bomber, the Canberra had a number of drawbacks when compared to most American aircraft in theater. First and foremost, the range accuracy of a level bomber was not necessarily as good as might be expected from a dive bomber, which meant that its employment in missions where friendly forces were in close proximity—especially troops-in-contact situations—needed to be considered carefully.[44] The Canberra's extended racetrack bombing patterns necessitated that artillery within a 9-mile (14.5-kilometer) radius of the target would have to be put on hold. The Canberra's straight 1,500-foot run-in to attack a target also made it potentially vulnerable to ground fire.[45] However, although level bombing had its limitations, it also proved to be extremely useful in certain situations. Indeed, Colonel Cregg Nolen Jr., the commander of the 35th TFW between January and July 1971, reflected at the end of his tour: "The level bombing capability of this aircraft complimented [sic] the dive deliveries of the fighters and provided excellent versatility for 7th A[ir] F[orce]."[46]

The Canberra's capacity to drop a stick of bombs (in addition to singly and in pairs) differentiated it from other aircraft in theater and offered a range of potential benefits, including the ability to vary the distance between individual bombs.[47] As a consequence, individual bombs could be spaced between 80 and 490 feet apart.[48] This capability proved to be particularly effective against "linear" targets such as trenches, tree lines, canals, and roads. It was estimated that during such operations the Canberra "could accomplish in one pass what other strike aircraft required up to six passes to achieve."[49] The key was matching suitable targets to the Canberra's bombing capabilities. One RAAF navigator has written of operations against targets along straight canals in the Mekong Delta: "It was a devastating sight to observe your first bomb strike a canal-side house and then see five neighbouring houses ('hooches') 75 metres apart all destroyed."[50] The ability to change fusing for each bombing run to either instantaneous or delayed provided even further flexibility.[51] Although it took some time for FACs to appreciate the Canberra's potential, they increasingly exploited the advantages it offered as they gained experience of operating with 2 Squadron.

The features of the operating environment in Vietnam gave the Canberra an opportunity to come into its own. The Canberra's capacity to bomb beneath low cloud cover was particularly useful during monsoon season.[52] Whereas dive bombers had difficulty identifying targets from 10,000 feet, the standard height from which their attacks started, the Canberra had the advantage of bombing from 3,000 feet and even as low as 1,000 feet under the right conditions.[53] One crew member recalled flying as low as 800 feet during one mission.[54] Flying Officer Richard O'Ferrall, a Canberra pilot, observed: "The Canberra could bomb under a low overcast, do it accurately, and was highly sought after by FACs during monsoonal weather in IV Corps for just this reason."[55] The Canberra's accuracy was significant, as General George S. Brown, commander, 7th Air Force, conceded in March 1970: "We are still not as effective in putting ordnance on target in weather conditions as we should be."[56] Air Commodore Neville McNamara, the commander of RAAF forces in Vietnam between 1971 and 1972, later recorded that the Canberra's "utility" was "highly valued by the US forces."[57]

The Canberra also had a distinct advantage in that its maximum of four hours of endurance were significantly more than the endurance of most tactical-strike aircraft in South Vietnam. The aircraft could loiter for around ninety minutes, as opposed to an average of approximately fifteen minutes for fighter aircraft.[58] As a consequence, the Canberras were regularly held up while other aircraft were brought in to attack a target.[59] Flying Officer David Robson, RAAF, a FAC with 19th Tactical Air Support Squadron, recalled: "We often had fighters stacked above the ground battle, and it was always an advantage to have extra loiter time over [the] target. We could also carry out an interim BDA and then use the Canberra to clean up."[60]

The greater loiter time also meant that the Canberras could await direction to new targets if original missions were cancelled as well as accept diverts anywhere throughout South Vietnam. Wing Commander John Whitehead, the commanding officer of 2 Squadron between November 1968 and November

1969, reflected: "The advantage of being able to hold overhead was that if the ground situation was changing in a particular area, we could wait to meet an emerging opportunity or even be deployed to another part of the country to carry out a more important bombing task."[61]

A subsidiary benefit for the RAAF crews was that because the Canberra was unique in theater, the Viet Cong and North Vietnamese Army labored under the premise that an aircraft represented a threat only if it were pointing in their direction, which meant that they often did not fire upon the Canberras. However, that theory was accurate only for dive bombers, and as the enemy forces discovered to their cost, by the time the aircraft were visible overhead, their bombs were only about three seconds behind. Nevertheless, concerns over enemy recognition of the Canberra as they became accustomed to its operations led to a decision on 15 November 1967 to remove the undersurface roundels from the aircraft's wings and reduce the fuselage roundels in size. Because antiaircraft fire was generally received only after aircraft had begun their attack dive, the Canberra's appearance was modified to increase its similarities with the B-57, which did not display underwing roundels or insignia.[62] Ultimately, although unfamiliarity was a recurring issue throughout 2 Squadron's commitment to Vietnam as incoming American personnel rotated into theater, the employment of different aircraft was beneficial as their capabilities proved to be complementary.

Familiarity Breeds Success

The utilization of unique RAAF aircraft inevitably created some teething problems when their strikes were controlled by FACs. The Australian FACs in theater had prior knowledge of the Canberra, but for American and, later, South Vietnamese FACs, there was somewhat of a learning curve as they adapted to working with level bombers.[63] FACs effectively needed to establish an attack heading that would often bring the Canberra on to the target on a pattern that ran along a linear feature. In part, this was easier to achieve for the Canberra because it was larger and slower than other jet aircraft and ultimately easier to keep in sight.[64] FACs could space two white phosphorous rockets in order to facilitate the Canberra's navigator in lining up the bombing run.[65] However, the target might need to be re-marked every time a bombing run was conducted because the smoke would disperse during the longer time required to prepare the Canberra for each sortie.[66]

Lieutenant Colonel Joe Madden, who served as the air liaison officer with the 1st Brigade, 4th US Infantry Division, claimed to be the first person to act as a FAC for a 2 Squadron Canberra. Madden recalled:

[W]e got real good work . . . during the battle of Dak To. We used them on mortar sites where I definitely knew where the mortar position was on the ground. The way we worked that, these guys needed forever. They would get off and then they would drive in like a normal, conventional bomber. They would have an IP [initial point] and then their release point. I worked out a deal with them. . . . In the 0–1 [Bird Dog] I only had four smoke rockets. . . . This guy wanted a mark down on the ground for an IP and then he wanted the target marked. So I said, "What I'll do, I'll drop a smoke can out for you as your IP and whenever you say you're lined up on the IP and you're a minute away from bombs away, you tell me, and then I'll mark the target for you." Well, he didn't like that idea, and I said, "Well, then, get the hell out of here. You aren't going to drop your bombs here if you don't do it that way." So he said, "Well, mate, I guess I can do it." . . . So I tossed the old smoke can out the window, and I said, "There's your IP." So he was running east and west as I recall. He drives in and he says, "60 seconds to go." So I roll in and zap a WILLY PETE [white phosphorous] on the ground, and he comes driving up and boy, that damn bomb hit right on the white smoke. There's no doubt about it. They're accurate as hell. So we did it that way. After the first bomb on the ground, it got smoke coming up, and that way then we didn't have to go through all this other foolishness of marking IP's and things like that. So they were very accurate.[67]

Like Madden, other American FACs seemed to enjoy working with the Canberras once they understood the aircraft's operating procedures. In reference to FAC knowledge of the aircraft's capability, Flying Officer Peter Nuske, a 2 Squadron pilot, opined: "They didn't understand what we could do initially, but once they had worked with the Canberra once or twice, they cottoned on and really made excellent use of the Canberra's advantages."[68] If a Canberra crew was provided with impractical instructions, then they simply advised the FAC how best to utilize the aircraft.[69]

Because FACs continually rotated in and out of theater, there was a diminishment in corporate memory, and an ongoing education process on the Canberra's capability was required. This education was achieved in part through 2 Squadron officers providing briefings at the USAF's so-called FAC University (504th Theater Indoctrination School) in Phan Rang. Furthermore, an exchange scheme was initiated that saw FACs and 2 Squadron crew obtain a perspective of how each other operated.[70] Flying Officer O'Ferrall confirmed that once FACs became accustomed to the aircraft, they "always adapted their strike plan and delivery profile requirements to best suit the Canberra's unique capabilities," and he concluded: "I never experienced a serious problem with a FAC, and even if they weren't used to us, they understood what we were going to [do] with little more than a sixty-second radio briefing."[71] Perhaps inevitably, in areas where 2 Squadron operated frequently, especially III and IV Corps, FACs had a good understanding of the Canberra's idiosyncrasies. However, when 2 Squadron was operating in I and II Corps, where the Canberra was not so well known, additional time was often required to establish operating procedures with the FACs.

For the Canberra crews, operations were simplified by the use of similar procedures by both American and Australian FACs.[72] In addition, some of the RAAF crews had prior experience working with USAF FACs.[73] However, despite the procedural similarities, accents proved to make operations challenging on occasion. During Flying Officer Errol McCormack's first flight using 2 Squadron's new call sign, an American radio operator was struggling to understand the signal "Magpie." The situation was clarified when an F-100 pilot from Phan Rang, who was used to the Australians, affirmed: "He's saying 'Magpie'—as in the bird. Tweet, tweet."[74] The problems posed by differing accents diminished as personnel became accustomed to operating together, although the rotation of personnel meant that the challenge recurred occasionally. The problems were generally minor and did not inhibit operations.

Things were not quite so easy during operations with South Vietnamese FACs. The combination of South Vietnamese FACs having generally "only a basic comprehension of the English language" and the Australians being "faster talking" created challenges.[75] Language difficulties meant that more time was frequently required to establish targets, and smoke rockets were sometimes fired before the Canberra was ready because the FAC responded when he heard key words in the planning process. The premature release of smoke was problematic due to the time required to prepare the bombing run

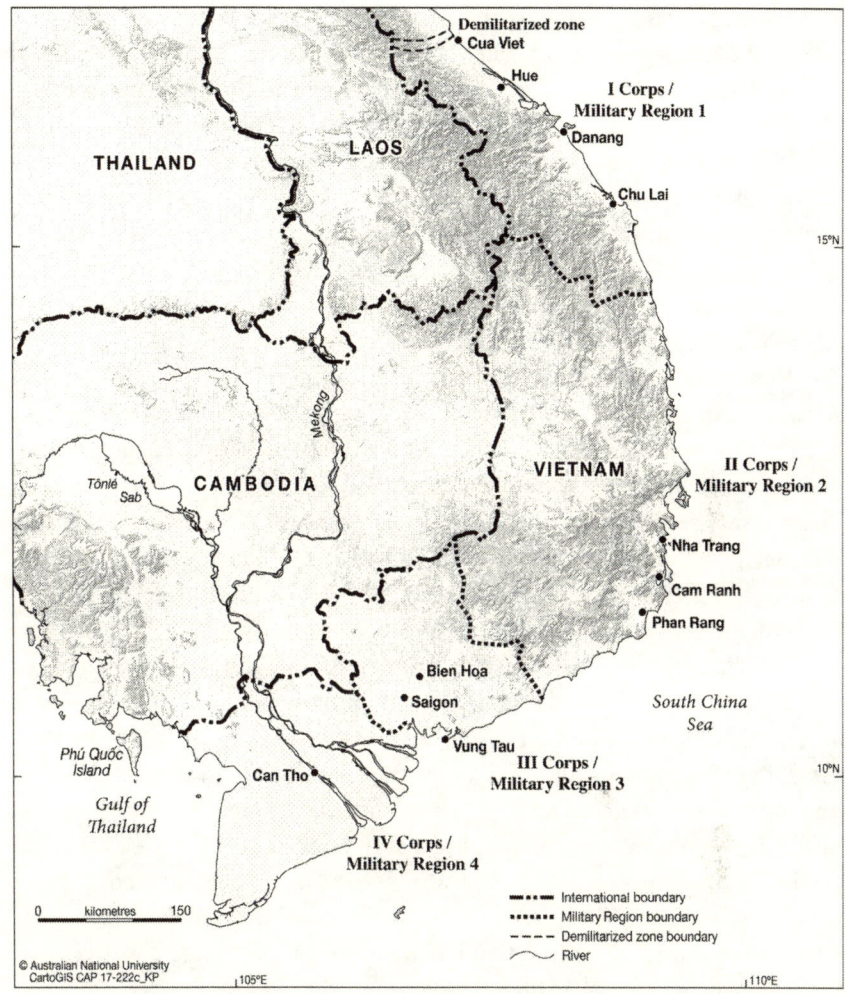

Military Regions of South Vietnam. The Corps Tactical Zones were renamed Military Regions on 1 August 1970. Map reproduced courtesy of the Australian National University, College of Asia and the Pacific (CAP), CartoGIS. © The Australian National University, CAP, CartoGIS.

and inevitably resulted in delays and occasionally aborted missions. Fortunately, the Canberra's longer loiter time meant that RAAF crews had more time to seek clarification than was the case for USAF fighter aircraft.[76]

As the number of missions with Vietnamese FACs increased, the exchange initiative was broadened in June 1970 to ensure greater understanding prior to the conduct of operations.[77] So-called combat pidgin English was also used to facilitate communications between Australian and South Vietnamese personnel.[78] As a consequence, a number of successful missions were conducted with South Vietnamese FACs. In one example, Pilot Officers John Kennedy and Alan Curr bombed an enemy position 5 miles (8 kilometers) southeast of Ben Tre, Kien Hoa Province. Six bombs were dropped 245 feet apart, destroying seventeen bunkers and twelve fortifications.[79] During another strike in support of friendly troops in contact, a 2 Squadron aircraft killed nine enemy soldiers and destroyed or damaged twelve bunkers and seven fortifications in Chuong Thien Province. The Canberra was diverted from its intended target and was directed by a South Vietnamese FAC to attack an enemy formation within 1,600 feet of friendly troops in what Pilot Officer Ross Hardcastle described as "a very good, well coordinated mission."[80]

Familiarization—ideally before operations but during if necessary—proved to be the keystone to making the multinational mix work. Education and familiarization efforts such as FAC University helped to ensure that the differences between aircraft and among techniques and procedures were understood and assisted in improving the effectiveness of multinational operations. Furthermore, the rapport that developed between FACs and the Canberra crews reassured the Australians that the targets provided to them were legitimate.[81] Liaison between Australian and American forces reinforced the strong relationship that had developed between the air forces and helped to ensure that any difficulties could be resolved promptly. Underpinning the RAAF–USAF relationship were a mutual respect and willingness to cooperate.

Playing by the Rules: Australian Rules of Engagement

The RAAF was limited to operations within South Vietnam throughout its involvement in the Vietnam War. Cambodia was considered to be a particularly sensitive area, with the Military Working Arrangement between the United States and Australia stating explicitly that "provisions will be made to insure [*sic*] that Australian Forces do not become involved in any incident along the Cambodian border, and in particular, that No 2 Squadron RAAF will not be allotted targets which may involve any danger of violation of the Cambodian border."[82] Number 2 Squadron was initially prevented from operating within 10 kilometers of the Cambodian border, but that restriction was later amended to 10 nautical miles (18.5 nautical kilometers) due to an "inadvertent change of the unit of measurement during correspondence."[83]

The limitations imposed on the RAAF necessitated that plans for the operation of 2 Squadron be changed. The MACV commander initially expressed the hope that, "if politically acceptable[,] RAAF squadron would also be employed to satisfy MACV requirements in Laos."[84] Although the restrictions, which were largely political in nature, were readily accepted by the 7th Air Force, they did cause an element of disappointment at the 35th TFW and among FACs.

After one successful Canberra strike against a base camp, bunker, and cave entrances near the Cambodian border on 17 December 1969, the FAC lamented having to target a "rock-pile" when there were numerous "juicy targets across the border."[85] Colonel Nolen described the restrictions as "unfortunate," observing of 2 Squadron: "On a daily basis of 12 sorties and 72 750 pound bombs, the effort may seem insignificant but taken in its entirety, much better effectiveness could have been achieved over the long run had the Canberras been utilized in Laos and Cambodia."[86] The restrictions became increasingly limiting as the focus of 35th TFW operations shifted. Colonel Walter Turnier, commander of the 35th TFW between June 1970 and January 1971, pointed out that there was "minimal commitment in-country" during his time in command, with the majority of sorties being flown over Laos and Cambodia.[87] The frustration was also shared by elements of 2 Squadron, with one Canberra navigator describing the ROE as "seemingly unrealistic" and "imposed by political constraints that impeded their operational effectiveness."[88] Notably, Wing Commander Evans was "embarrassed" by not being able to operate over Cambodia and was compelled—to no avail—to write to the Chief of the Air Staff to inform him that 2 Squadron could be effective there, particularly at night.[89]

Although the restrictions were primarily political, the prohibition of operations over North Vietnam also had a very practical purpose. Wing Commander Whitehead has noted: "The Canberra was not properly equipped with ECM [electronic countermeasures] and could not compare with the

USAF ground attack aircraft that were much faster and more manoeuvrable in low-level operations."[90] Another RAAF pilot opined: "The Squadron would have been annihilated within a matter of weeks due to the ferocious enemy opposition," which consisted of MiG aircraft, surface-to-air missiles, and antiaircraft batteries.[91] The risk was emphasized when an RAAF Canberra was shot down by a surface-to-air missile on 14 March 1971 while operating near the demilitarized zone.[92] Wing Commander Evans, in spite of the risks, favored conducting a single mission over North Vietnam, with the support of American aircraft utilizing jamming equipment. Although acknowledging that such a mission would have been "dicey," Evans believed that it would have been feasible at least once out of surprise, and he felt that it would "show our [2 Squadron's] willingness and ability to do it."[93] Despite Wing Commander Evans's belief that an RAAF operation over North Vietnam would have enhanced the reputation of both 2 Squadron and Australia more broadly, permission was never granted for it.

Air Commodore C. H. Spurgeon, who commanded the RAAF in Vietnam between 1970 and 1971, recalled that the tasking of aircraft created problems due to misunderstandings about the nature of the Canberras: "The targeting and mission responsibility for those aircraft was totally in the hands of a USAF operations room run by a lieutenant colonel. Our Canberras were listed on the mission board as B-57s, when in fact in some critical aspects they were quite different."[94] Whitehead noted that "one or two missions" where the 7th Air Force targeting center tasked 2 Squadron for high-level Combat Skyspot sorties were denied due to the potential to overfly the border.[95] However, missions could easily be adapted at the planning stage, and apart from some potential frustration on both sides, it was not onerous to task an alternative American squadron. Things were somewhat more difficult when FACs requested RAAF Canberras to divert to unplanned missions that would have contravened their ROE. One pilot remembered: "The FAC (who would probably be unaware of our restrictions) would be quite frustrated (as well as us) that we could not comply."[96]

Although the restrictions were outlined clearly, adherence to those constraints could be more difficult in practice. First and foremost, the border was not "neatly marked by a white painted picket fence," and it was not impossible for aircraft to cross the border without realizing it.[97] Moreover, the navigational aids in the Canberra were not considered reliable at low level.[98] As one Australian outlined, "If you are busy manoeuvring at

8 miles [12.9 kilometers] a minute at night, high over featureless, unpopulated terrain while talking to Combat Skyspot, it would be near impossible to be certain where you are in relation to an unmarked border."[99]

Therefore, although crews were conscious of the need to observe the restrictions rigidly, the potential for unintended breaches was a reality of the nature of the operating environment and the limitations of the technology onboard the aircraft. One analyst has since observed: "The inviolability of borders was a legitimate political constraint to set on our forces, but in some circumstances it was quite impractical."[100] Indeed, since the war, it has been reported but not confirmed officially that in exceptional circumstances, such as troops in contact, occasional breaches may have occurred at the request of FACs.[101] The Australian crews were in an incredibly difficult position when the lives of their allies were at risk on the ground but political constraints militated against their involvement. Constraints were accepted and understood at the strategic level, but adherence to them was challenging at the tactical level.

LOGISTICS: SUSTAINING OPERATIONS

As in any military operation, but especially those of a multinational nature, logistics was an important factor in the successful integration of 2 Squadron during the Vietnam War. The facilities provided at Phan Rang were essential for the sustainment of 2 Squadron operations across the course of its deployment. Stephens has stressed: "It would be naive to think that the RAAF could have deployed to Vietnam in the strength that it did for as long as it did had those amenities not been provided by the Americans."[102] Although US support was vital, the Australians' willingness to pay their own way was important from the perspective of both national pride and relations at the strategic level. The existence of a financial working arrangement that outlined clearly the basis for reimbursement was an important step.[103] However, the agreement reached between the commander of Australian Force Vietnam and the MACV commander was not well known, and Wing Commander Evans recorded that he was required to physically show the document to the commanding officer of the 35th TFW to correct the notion that the Australians were free riding. He recalled: "The Wing Commander was astounded when he saw that we were paying for everything; our rations, our fuel, and bombs that we needed to acquire from the USAF. We were much better serviced and

appreciated once this was established."[104] Although the Americans were notoriously generous in providing support, the existence of the financial working
arrangement helped to facilitate relations and smooth the support process.

The nature of 2 Squadron operations meant that the provision of muni-
tions was a key issue that required immediate resolution. The RAAF hoped
to make use of the large number of British-pattern bombs left over from the
Second World War, but due to a lack of storage space the Americans rejected
an initial plan to store an eighteen-month supply of 500-pound bombs in
Vietnam.[105] The USAF could provide space for only a forty-five-day supply at
Phan Rang, with a reserve area at Cam Ranh Bay for backup supplies.[106]
Although 27,568 bombs were scheduled for shipment from Australia to Vietnam, it was hoped that American bombs could be provided to the RAAF in
the short term. However, in an early demonstration of the utility of standardization, it was discovered that the RAAF aircraft would need to be modified
and crews retrained to use American bombs.[107]

The use of the British-pattern bombs created a number of practical challenges. The 500-pound and 1,000-pound bombs had varying fusing arrangements and tails that affected the ballistic characteristics. Wing Commander
Vin Hill noted that "life was simpler when the last of the old Australian stocks
were used up," but the variations could be accommodated, although they
had to be communicated to the Skyspot controller.[108] Problems with duds
and late release created further concerns. As a result, it was hoped that the use
of US munitions might simplify operations. The USAF agreed to provide
M117 bombs and associated components to the RAAF from US sources, with
discrepancies in consignments being rectified by in-theater stocks.[109]

The changeover to US M117 750-pound bombs took place on 11 August
1968 following months of preparatory work and trials, although 1,000-pound
bombs were reintroduced between November 1969 and May 1970 following
the discovery of additional stocks in Darwin.[110] Modifications were made to
the aircraft, including the fitment of an adaptor rack, and 2 Squadron dropped
multiple loads of M117 bombs in March 1968.[111] Wing Commander Evans
recorded that the trial was "successful" and that "no difficulty in converting
to exclusive use of these weapons is expected."[112] Subsequent carriage and
ballistic checks confirmed the initial findings.[113] In June 1968, seven sorties
were conducted daily over the course of a week using M117 bombs during both Skyspot and daylight visual operations.[114] Although the use of
American bombs required some modification, it was a minor inconvenience

in comparison to the capacity to rely on US sources for replenishment. Compatibility increased efficiency, but the benefits of employing a range of munitions was demonstrated by the use of 1,000-pound bombs for particular roles, such as creating clearings in the jungle.[115]

Aside from the anticipated support provided in things such as fuel and munitions, the conduct of 2 Squadron operations in South Vietnam and alongside the USAF necessitated the provision of additional equipment, which was readily supplied from American sources. Skyspot operations required radar transponders to be fitted to the Canberra. Initial delays in obtaining the transponder beacons limited the effectiveness of early operations, as did the fitment of them in the cockpit.[116] However, 2 Squadron had been provided with seven transponder beacons by June 1967 and moved them to the tail cone, which reduced the shielding effect on emissions and improved effectiveness.[117] When availability issues recurred, leading to only three transponder beacons for 2 Squadron in September 1967, it was noted that the shortage was "common to all squadrons of 35th TFW" and that the RAAF had a "fair proportion of the transponders available to the Wing."[118] The preponderance of Skyspot operations ensured that the USAF's willingness to provide the transponder beacons was essential for 2 Squadron.

The USAF also provided equipment to enhance the safety of RAAF crews. RT-10 survival radios were supplied to 2 Squadron in sufficient numbers for one to be provided to each crew member to increase the chance of downed personnel being recovered.[119] It was noted that "US pilots recognize the RT-10 transmission more readily, and are unwilling to investigate unfamiliar transmissions due to known enemy tactics of setting up decoy beacons."[120] Tree-escape kits and LCU 11/P tree-lowering devices were also acquired from the USAF.[121] In sum, the provision of equipment from USAF sources increased the efficiency of 2 Squadron operations and enhanced the crews' safety. The USAF's generosity in sharing equipment in conjunction with the RAAF's willingness to pay its own way increased the efficiency of 2 Squadron's operations without creating an undue strain on either partner.

CONCLUSION: CONTEXTUALIZING THE AUSTRALIAN EXPERIENCE

In reflecting on 2 Squadron's achievements, Flight Lieutenant Robert Howe, a Canberra navigator, concluded: "I believe that the US military got the maximum service out of 2 Squadron Canberras by judicious use of its effective capabilities while avoiding being prisoner to its limitations."[122] The concept of maximizing capabilities and minimizing limitations was crucial to the successful integration of the RAAF and the USAF. Straightforward command-and-control arrangements smoothed integration between the two air forces, and the process was eased by the fact that 2 Squadron was a self-contained unit, thereby avoiding a number of potential points of friction. In addition, as the RAAF made the 7th Air Force fully aware of the clear and consistent restrictions on 2 Squadron operations, the variation with US ROE did not create major problems. However, immediate requests were more problematic. It was testament to the relationship between the RAAF and the USAF that the complications created through the utilization of differing ROE did not undermine Australian involvement in air operations or diminish 2 Squadron's reputation.

The Vietnam War also highlighted that although there are a number of advantages to standardization, not least when it comes to logistics, there are also a range of potential benefits to operating a variety of platforms. Although the Canberra was unsuitable for some roles, it offered capabilities different from other aircraft in theater, which served as a force multiplier. One Australian source later observed: "The lessons [of Vietnam] appear to be that years of combined training and standardization in peace are essential if there is to be rapid, effective battlefield cooperation."[123] In the absence of standardized platforms and with only limited training opportunities, the air forces had to learn how to cooperate effectively through experience accumulated during the course of operations. Although it was not an ideal solution, it did lead to an effective relationship and ultimately a commendable performance. Brigadier General W. T. Galligan, who commanded the 35th TFW between August 1969 and June 1970, reflected on 2 Squadron: "I can't speak highly enough of their outstanding professionalism, across the board. I only wish that all USAF units could do as well."[124] For 2 Squadron, Vietnam operations were a "professionally rewarding" experience.[125] Irrespective of the outcome of the conflict, Air Commodore Gary Waters, RAAF, has assessed from a national perspective that the "operational skills forged in Vietnam provided the foundation for the RAAF for the next 20 years."[126]

In addition, the experiences of the Vietnam War enhanced RAAF–USAF relations. The strength of that existing relationship undoubtedly facilitated cooperation and minimized the friction experienced at the start of operations. Nevertheless, the use of nonstandardized aircraft and the imposition of different ROE meant that a learning curve was inevitable. The RAAF's role as a willing and efficient ally brought it respect and understanding, which helped to underpin its relationship with the US 7th Air Force. The successful cohesion between the RAAF and the USAF was unable to influence the outcome of the conflict, but it did provide a platform for future cooperation in both peace and war.

NOTES

1. Jonathan Colman and J. J. Widen, "The Johnson Administration and the Recruitment of Allies in Vietnam, 1964–1968," *History* 94, no. 4 (2009): 485.

2. There are, however, some excellent works on the RAAF's involvement in the Vietnam War, including the official history by Chris Coulthard-Clark, *The RAAF in Vietnam: Australian Air Involvement in the Vietnam War, 1962–1975* (Sydney: Allen & Unwin, 1995). In addition, there are important works on the involvement of particular squadrons. For 2 Squadron operations, for example, see John Bennett, *Highest Traditions: The History of No 2 Squadron, RAAF* (Canberra: Australian Government Publishing Service, 1995), and Bob Howe, *Dreadful Lady over the Mekong Delta: An Analysis of RAAF Canberra Operations in the Vietnam War* (Canberra: Air Power Development Centre, 2016).

3. David Wilson, "C130 Operations, No. 5 Airfield Construction Squadron, Airfield Defence, Logistics, Photographic Interpreters," in *The RAAF in the War in Vietnam,* ed. John Mordike (Canberra: Air Power Studies Centre, 1999), 83–93; Graham O'Brien, *Always There: A History of Air Force Combat Support* (Canberra: Air Power Development Centre, 2009), 67–72.

4. RAAF personnel also saw service as photographic interpreters with the US Air Force's 6470th Reconnaissance Technical Squadron and 460th Reconnaissance Wing and as radar controllers.

5. In 1968, for example, 2 Squadron accounted for 16 to 20 percent of the 35th TFW's bomb damage, although flying just 5 percent of the sorties. See Mr. James T. Bear, HQ Pacific Air Forces, Directorate, Tactical Evaluation, Contemporary Historical Examination of Current Operations (CHECO) Division, "Project CHECO Southeast Asia Report: The RAAF in SEA," 30 September 1970, 17, 19, collection 372, box 207, Naval History and Heritage Command (NHHC), Washington, DC.

6. Combined Review Team, *Review of Strategic Level Interoperability between the Military Forces of Australia and the United States of America* (Canberra: Australian Department of Defence, October 2002), 62.

7. Alan Ryan, *From Desert Storm to East Timor: Australia, the Asia-Pacific, and the "New Age" Coalition Operations* (Canberra: Land Warfare Studies Centre, January 2000), 19.

8. Quoted in Christopher Hubbard, *Australian and US Military Cooperation: Fighting Common Enemies* (Aldershot, UK: Ashgate, 2005), 24.

9. Despina Tramoundanis, *Australian Air Power in Joint Operations* (Canberra: Air Power Studies Centre, 1995), 4.

10. Group Captain L. H. Williamson for Chief of the Air Staff, Department of Air Organization Directive No. 26/68, "Organization and Administration of the RAAF Units of the Australian Force Vietnam," 15 October 1968, A703, 635/2/566 Part 3, National Archives of Australia (NAA), Canberra.

11. Air Commodore N. Ford, Headquarters RAAF Butterworth, "Administrative Order No. 3/67," 30 March 1967, RAAF, No. 2 Squadron, Miscellaneous, 1967, Office of Air Force History, Australia (OAFHA), Canberra; "Unit History Sheet 1967," April 1967, RAAF, No. 2 Squadron, OAFHA; "Memorandum of Understanding between Commander 7th Air Force and Deputy Chief Air Staff Royal Australian Air Force," 5 January 1967, Record Group (RG) 472, box 3, Records of United States Forces in Southeast Asia, National Archives and Records Administration (NARA), College Park, MD. Number 35 Squadron had also been placed under the operational control of the 7th Air Force (David Horner, *Australian Higher Command in the Vietnam War* [Canberra: Australian National University Press, 1986], 22).

12. "Commanding Officer's Report, No. 2 Squadron, April 1967," 10 May 1967, OAFHA.

13. Weekly Report for the President, "Seventh Air Force Will Give Logistical Support to Royal Australian Air Force (RAAF) Canberra Squadron Recently Deployed to Phan Rang AB, Republic of Vietnam (RVN)," 11 April 1967, RG 472, box 11, Records of United States Forces in Southeast Asia, NARA; Group Captain L. H. Williamson for Chief of the Air Staff, Department of Air Organization Directive No. 26/68, "Organization and Administration of the RAAF Units of the Australian Force Vietnam," 15 October 1968, A703, 635/2/566 Part 3, NAA.

14. Chris Clark, "A History of Australian Air Power and Irregular Warfare," in *The Art of Air Power,* ed. Wing Commander Keith Brent (Canberra: Air Power Development Centre, 2011), 157.

15. MACJ312 to Commander in Chief, Pacific Fleet (CINCPAC), "RAAF Canberra Squadron," 2 January 1967, RG 472, box 11, Records of United States Forces in Southeast Asia, NARA.

16. David Evans, *Down to Earth: The Autobiography of Air Marshall David Evans, AC, DSO, AFC* (Canberra: Air Power Development Centre, 2011), 147.

17. Alan Stephens, "Observations on an Expeditionary War of Choice: The RAAF in Vietnam, 1964–1971," in *Air Expeditionary Operations from World War II until Today,* ed. Keith Brent (Canberra: Air Power Development Centre, 2009), 54.

18. Brigadier General Frank Gailer Jr., written interview by the author, 30 August 2012. In the chapter text, interviewees are referred to by their rank at the time of their deployment to Vietnam, but their final rank is given in the notes.

19. Evans, *Down to Earth,* 148.

20. Ian McNeill has observed that "Australian airmen had already operated as part of a higher-level U.S. formation during the Korean War" ("The Australian Army and the Vietnam War," in *Australia's Vietnam War,* ed. Jeff Doyle, Jeffrey Grey, and Peter Pierce [College Station: Texas A&M University Press, 2002], 49).

21. Air Commodore John Whitehead, written interview by the author, 10 February 2012.

22. Squadron Leader Richard O'Ferrall, written interview by the author, 23 December 2011.

23. Richard A. Cody and Robert L. Maginnis, "Coalition Interoperability: ABCA's New Focus," *Canadian Army Journal* 10, no. 1 (Spring 2007): 87.

24. Roger Hilsman to David Ball, Administrator, Agency for International Development, "1550 Determination for Financial Assistance to Australia for the Procurement of a Third Charles F. Adams Class Destroyer in the United States," memorandum from FE, 18 April 1963, RG 59, box 16, NARA.

25. "Minutes of Australian Mission Meeting, 6th June 1957," 20 June 1957, RG 59, box 2, NARA.

26. Stephan Frühling, "Australian Strategic Guidance since the Second World War," in *A History of Australian Strategic Policy since 1945,* ed. Stephan Frühling (Canberra: Defence Publishing Service, 2009), 21.

27. Ian Bellamy and James Richardson, "Australian Defence Procurement," in *Problems of Australian Defence,* ed. H. G. Gelber (Melbourne: Oxford University Press, 1970), 257.

28. Edward Clark, Ambassador, Embassy of the United States, Canberra, to Mr. Robert S. Lindquist, Country Director, Australia, New Zealand, and South Pacific Islands Affairs, Department of State, 26 September 1966, RG 59, box 9, NARA.

29. For more on the RAN's experience alongside the USN in Vietnam, see Steven Paget, "On a New Bearing: The Re-organised Royal Australian Navy at War in Vietnam," *The Mariner's Mirror* 101, no. 3 (2015): 283–303. For more on the F-111, see Mark Lax, *From Controversy to Cutting Edge: A History of the F-111 in Australian Service* (Canberra: Air Power Development Centre, 2010).

30. The Australian government declined a US offer to lend two squadrons of B-47s due to "the very considerable disturbance which would be involved to the RAAF . . . and the limited period during which the aircraft would be operational with us." It was also noted that "the safe fatigue life of the Canberra could favourably cover the presently estimated period required for the introduction of the F-111A aircraft into squadron service" (US Department of State, "Department of Defense for McNamara from Battle," incoming telegram, 3 June 1964, RG 59, box 5, NARA).

31. US Defense Attaché Office Canberra to RUEPJS/DIA, "Augmentation in Vietnam," memorandum, 4 January 1967, RG 472, box 11, Records of United States Forces in Southeast Asia, NARA.

32. Chris Coulthard-Clark, "The Air War in Vietnam: Re-evaluating Failure," in *The War in the Air: 1914–1994,* ed. Alan Stephens (Canberra: Air Power Development Centre, 2009), 118.

33. Quoted in Coulthard-Clark, *The RAAF in Vietnam,* 95.

34. Presentation by Commander, Australian Force Vietnam, "The Situation in South Vietnam and Australian Participation," CGS Exercise 1970, AWM 98, R841/2/7, Australian War Memorial, Canberra. It should be noted that B-52 aircraft were also used in a tactical role.

35. Wing Commander V. J. Hill, "No. 2 Squadron Operations—South Vietnam: The First Month," 20 May 1967, OAFHA. At the outset of the first rotation, each crew member of 2 Squadron went for a familiarization flight in an F-100 (Air Marshal Errol McCormack, interview by the author, Canberra, 13 February 2012).

36. Coulthard-Clark, *The RAAF in Vietnam,* 191.

37. Bear, "Project CHECO Southeast Asia Report," 30 September 1970, 19.

38. Ibid., 20.

39. Royal Australian Air Force, No. 2 Squadron, "Unit History Sheet 1967," June 1967, OAFHA.

40. Wing Commander R. B. Aronsen, "Report for the Month of September 1967 on the Activities of the RAAF Force—Vietnam," 18 October 1967, A703, 580/1/277 Part 3, NAA.

41. Evans, *Down to Earth,* 148.

42. The shift had a significant effect on morale as Skyspot operations were "routine and boring." Alan Stephens notes: "Some crews were flying 90 per cent of their missions at night, with little required from the pilot and navigator other than accurate flying and correct weapons selection" (*The Royal Australian Air Force: A History* [Melbourne: Oxford University Press, 2006], 272).

43. Whitehead, written interview.

44. Group Captain John Tyrrell, written interview by the author, 24 January 2012.

45. Wing Commander Brian Fooks, written interview by the author, 4 March 2012.

46. Colonel Cregg P. Nolen Jr., "Project Corona Harvest End-of-Tour Report," n.d., call no. K740.131, IRIS no. 00524473, Office of Air Force History, Maxwell Air Force Base (OAFH), AL.

47. Wing Commander S. D. Evans, "Commanding Officer's Report, No. 2 Squadron, December 1967," 5 January 1968, A703, 580/1/277 Part 3, NAA.

48. Coulthard-Clark, *The RAAF in Vietnam,* 194.

49. Bear, "Project CHECO Southeast Asia Report," 30 September 1970, 18.

50. Wing Commander Robert Howe, written interview by the author, 4 January 2012.

51. Tyrrell, interview.

52. The operations built on the experience of conducting low-level visual bombing during the Malayan Emergency (Alan Stephens, *Going Solo: The Royal Australian Air Force, 1946–1971* [Canberra: Australian Government Publishing Service, 1995], 303).

53. Coulthard-Clark, *The RAAF in Vietnam,* 193–94.

54. T. E. Bell, *B-57 Canberra Units of the Vietnam War* (Oxford: Osprey, 2011), 76.

55. O'Ferrall, interview.

56. "Interview with General George S. Brown, CMDR, 7th AF, by Mr Kenneth Sams and Lt. Col. Richard Kott, 30 March 1970," revised 2 April 1970, RG 472, box 246, Records of United States Forces in Southeast Asia, NARA.

57. Neville McNamara, *The Quiet Man: The Autobiography of Air Chief Marshal Sir Neville McNamara* (Canberra: Air Power Development Centre, 2005), 150.

58. Coulthard-Clark, *The RAAF in Vietnam,* 194–95.

59. Wing Commander S. D. Evans, "Commanding Officer's Report, No. 2 Squadron, March 1968," 15 April 1968, A703, 580/1/277 Part 4, NAA; O'Ferrall, interview.

60. Wing Commander David Robson, written interview by the author, 7 January 2012.

61. Whitehead, interview.

62. Air Commodore J. F. Lush, "Commanding Officer's Report, No. 2 Squadron, November 1967," 29 November 1967, A703, 580/1/277 Part 3, NAA.

63. That prior knowledge shaped the FACs' views occasionally. Although acknowledging that the Canberras were "operated by very professional crews," Flight Lieutenant Garry Cooper felt that the aircraft's characteristics were "more suited to saturation bombing," and due to the Canberras' shortcomings he "did not like using them in contacts close to

friendly troops" (Garry Cooper, *Sock it to 'Em Baby: Forward Air Controller in Vietnam* [Sydney: Allen & Unwin, 2006], 69). It was unfair to prejudge the aircraft, though, because extensive training in it led to a high degree of accuracy. During one mission in support of the US 25th Infantry Division, which had called for close air support while sweeping a village on 27 July 1969, the FAC and ground commander communicated: "We look forward to working with the Magpies because of such pinpoint accuracy" (quoted in Bennett, *Highest Traditions*, 313).

64. Fooks, interview.

65. Robson, interview.

66. Coulthard-Clark, *The RAAF in Vietnam*, 195.

67. Lieutenant Colonel Joe Madden, interview by Lieutenant Colonel V. H. Gallacher and Major Lyn R. Officer, US Air Force Oral History Program, interview 654, 7 February 1973, call no. K239.0512-654, IRIS no. 00904723, OAFH.

68. Flight Lieutenant Peter Nuske, written interview by the author, 9 January 2012.

69. Air Marshal David Evans, interview by the author, Canberra, 24 November 2011.

70. Howe, interview.

71. O'Ferrall, interview.

72. Tyrrell, interview. The RAAF had importantly obtained experience of attacking targets marked by Royal Air Force Auster aircraft during the Malayan Emergency (Steve Eather, *Odd Jobs: RAAF Operations in Japan, the Berlin Airlift, Korea, Malaya, and Malta, 1946-1960* [Point Cook, Australia: Royal Air Force Museum, 1996], 53-54).

73. Some RAAF FACs (such as Flying Officer Robson) had similarly gained experience of operating with the USAF before their deployment to Vietnam (Robson, interview).

74. McCormack, interview.

75. Howe, *Dreadful Lady over the Mekong Delta*, 84.

76. Bennett, *Highest Traditions*, 322.

77. Howe, interview.

78. Howe, *Dreadful Lady over the Mekong Delta*, 75.

79. Directorate of Information, Headquarters 7th Air Force, "United States Air Force Daily Summary, Sunday, Jan. 18, 1970," news release, n.d., call no. K740.951-52, IRIS no. 00527942, OAFH.

80. Directorate of Information, Headquarters 7th Air Force, "United States Air Force Daily Summary, Friday, Aug. 7, 1970," news release, n.d., call no. K740.951-52, IRIS no. 00527949, OAFH.

81. Howe, *Dreadful Lady over the Mekong Delta*, 85.

82. "Military Working Arrangement between Commander, United States Military Assistance Command Vietnam, and Chairman, Chiefs of Staff Committee Australia," 30 November 1967, RG 472, box 3, Records of United States Forces in Southeast Asia, NARA.

83. Coulthard-Clark, *The RAAF in Vietnam*, 281.

84. COMUSMACV [Commander, MACV] to CINCPAC, "Australian Acft Sqdn for SVN (U)," memorandum, 22 April 1966, RG 472, box 36, Records of United States Forces in Southeast Asia, NARA.

85. Quoted in Howe, *Dreadful Lady over the Mekong Delta*, 146.

86. Nolen, "Project Corona Harvest End-of-Tour Report."

87. Colonel Walter C. Turnier, "Project Corona Harvest End-of-Tour Report," n.d., call no. K740.131, IRIS no. 00524585, OAFH.

88. Quoted in Howe, *Dreadful Lady over the Mekong Delta,* 166.

89. Evans, interview.

90. Whitehead, interview.

91. Nuske, interview. American aircraft losses demonstrated the validity of Australian concerns, with 777 being shot down over North Vietnam between 1965 and 1967 (Mark Clodfelter, *The Limits of Air Power: The American Bombing of North Vietnam* [Lincoln: University of Nebraska Press, 2006], 131).

92. Robert Jackson, *Canberra: The Operational Record* (Washington, DC: Smithsonian Institution Press, 1989), 92.

93. Evans, interview.

94. C. H. Spurgeon, "Discussion," in *The War in the Air,* ed. Stephens, 123–24. As a result of the confusion that sometimes originated from the use of similar but different aircraft, Howe has opined that the RAAF "erred in not taking up a permanent position within the [Tactical Air Control Center]" (*Dreadful Lady over the Mekong Delta,* 65).

95. Whitehead, interview.

96. Nuske, interview.

97. Spurgeon, "Discussion," 123–24.

98. Whitehead, interview.

99. Tyrrell, interview.

100. Chris Coulthard-Clark, "Discussion," in *The War in the Air,* ed. Stephens, 124.

101. One USAF FAC reported directing RAAF Canberras to strike a bridge in southern Laos, which they successfully destroyed (Bell, *B-57 Canberra Units of the Vietnam War,* 76–79).

102. Stephens, "Observations on an Expeditionary War of Choice," 49–50.

103. For example, the USAF was reimbursed for munitions at cost plus 20 percent for accessorial charges ("Memorandum of Understanding between Commander 7th Air Force and Deputy Chief Air Staff Royal Australian Air Force," 5 January 1967).

104. Evans, *Down to Earth,* 147.

105. Coulthard-Clark, *The RAAF in Vietnam,* 98.

106. "Memorandum of Understanding between Commander 7th Air Force and Deputy Chief Air Staff Royal Australian Air Force," 5 January 1967.

107. Coulthard-Clark, *The RAAF in Vietnam,* 98, 184.

108. Howe, *Dreadful Lady over the Mekong Delta,* 111; Coulthard-Clark, *The RAAF in Vietnam,* 190–91. Conversely, while labelling the 500-pound bombs as "useless," Wing Commander Evans described the 1,000-pound bombs as "wonderful . . . probably the best bombs in Vietnam," and added that he was "very disappointed" when 2 Squadron had to switch to the 750-pound US bombs (Evans, interview).

109. "Memorandum of Understanding between Commander 7th Air Force and Deputy Chief Air Staff Royal Australian Air Force," 5 January 1967; Wing Commander A. J. Read, "Monthly Report—July 1968 by Staff Officer Equipment," 15 August 1968, A703, 580/1/277 Part 4, NAA.

110. Royal Australian Air Force, No. 2 Squadron, "Unit History Sheet 1968," August 1968, OAFHA.

111. Wing Commander E. Harrison, "Monthly Report—March 1968 by Staff Officer Engineering," 15 April 1968, A703, 580/1/277 Part 4, NAA.

112. Wing Commander S. D. Evans, "Commanding Officer's Report, No. 2 Squadron, March 1968," 15 April 1968, A703, 580/1/277 Part 4, NAA.

113. Wing Commander S. D. Evans, "Commanding Officer's Report, No. 2 Squadron, April 1968," 13 May 1968, A703, 580/1/277 Part 4, NAA.

114. Wing Commander S. D. Evans, "Commanding Officer's Report, No. 2 Squadron, June 1968," 18 July 1968, A703, 580/1/277 Part 4, NAA.

115. Coulthard-Clark, *The RAAF in Vietnam,* 197.

116. "Commanding Officer's Report, No. 2 Squadron, April 1967," 10 May 1967, OAFHA; Coulthard-Clark, *The RAAF in Vietnam,* 190.

117. "Commanding Officer's Report, No. 2 Squadron, June 1967," 3 July 1967, OAFHA.

118. "Commanding Officer's Report, No. 2 Squadron, September 1967," 5 October 1967, OAFHA.

119. Wing Commander S. D. Evans, "Commanding Officer's Report, No. 2 Squadron, May 1968," 18 June 1968, A703, 580/1/277 Part 4, NAA.

120. Headquarters RAAF Element HQAFV, "Report for the Month of April 1968 on the Activities of RAAF Force Vietnam," 13 May 1968, A703, 580/1/277 Part 4, NAA.

121. Wing Commander S. D. Evans, "Commanding Officer's Report, No. 2 Squadron, July 1968," 15 August 1968, A703, 580/1/277 Part 4, NAA; Evans, "Commanding Officer's Report, No. 2 Squadron, June 1968," 18 July 1968.

122. Howe, interview.

123. Quoted in Thomas-Durell Young, *Australian, New Zealand, and United States Security Relations, 1951–1986* (Boulder, CO: Westview Press, 1992), 121.

124. Quoted in Bear, "Project CHECO Southeast Asia Report," 30 September 1970, 16–17.

125. Alan Stephens, *Power Plus Attitude: Ideas, Strategy, and Doctrine in the Royal Australian Air Force, 1921–1991* (Canberra: Australian Government Publishing Service, 1992), 124.

126. Air Commodore Gary Waters, "One Hundred Years of Air Power—an Australian Perspective," in *100 Years of Aviation: The Australian Military Experience,* ed. Wing Commander Keith Brent (Canberra: Royal Australian Air Force Aerospace Centre, 2004), 196.

Coalition Air Power in the Gulf War, 1991

Richard P. Hallion

Less than three months after the end of the Gulf War of 1991, the noted air power analyst Air Vice Marshal R. A. "Tony" Mason, Royal Air Force (RAF), judged it "the apotheosis of twentieth-century air power."[1] In thirty-nine days of round-the-clock strategic, operational, and tactical air and cruise-missile attacks, followed by a four-day air-and-artillery-supported armored mechanized assault, a coalition mandated by the United Nations (UN) liberated Kuwait and, striking deep into Iraq itself, occupied 28,500 square miles (73,815 square kilometers) of territory, taking more than 87,000 prisoners. In a furious welter of explosive and cacophonous destruction, it broke the world's sixth-largest air force and reduced the world's fourth-largest army to merely the fourth-largest army inside Iraq.[2] The outcome confounded pre-war prognosticators, shattered conventional wisdom about air campaigns and cross-border maneuver warfare, led to widespread reform of military doctrine and forces, and triggered a still unresolved debate about whether it constituted a transformational "Revolution in Military Affairs" or simply a single uniquely situational (and thus anomalous) event.[3]

Though video of high-tech F-117 "stealth" fighters, Patriot missiles, Apache gunships, and "smart" bombs dominated the media, the outcome represented the collective achievement of coalition partners that planned and fought together to liberate Kuwait from Saddam Hussein. A brief summation of what they brought—and the costs they bore in both material and human terms—offers ample measure of that contribution. Excluding the United States, coalition members

- furnished all bases, airfields, camps, operating locations, and ports;
- furnished more than 90 percent of fuel used in Desert Shield/ Storm;
- furnished a third of all troops and other military personnel;
- furnished a quarter of all fixed-wing aircraft (including almost a third of all fighter/attack aircraft and almost a third of all airlifters);
- furnished nearly half of all naval vessels;
- furnished more than a third of all armored and mechanized vehicles;
- offset in cash or in kind nearly 90 percent of US incremental crisis and war costs;
- suffered a quarter of all aircraft losses from Iraqi fire;
- suffered more than a third of all combat deaths from Iraqi fire;
- suffered more than half of all combat wounded from Iraqi fire.

Quite clearly, then, coalition members more than pulled their weight in influencing the war's outcome.

FROM CRISIS ONSET THROUGH DESERT SHIELD

Saudi Arabia's King Fahd and other Persian Gulf leaders saw Iraq's invasion of Kuwait in the early hours of 2 August 1990 as a perfidious, dishonorable, and ungrateful act of betrayal—not merely of Kuwait and the Gulf Cooperation Council but also of the Arab–Islamic world as a whole. European and American leaders were likewise shocked and angered, with Britain's prime minister Margaret Thatcher and France's president François Mitterrand reacting most strongly.

On 7 August, US Central Command initiated force deployment Operation Plan 1002-90, marking the beginning of Operation Desert Shield, and Prime Minister Thatcher ordered the RAF to send strike aircraft to the Gulf, marking the beginnings of Operation Granby, the British effort in the Gulf War. The first American aircraft—twenty-three F-15Cs for Saudi air defense—arrived at Dhahran on 8 August, and the first British aircraft—twenty-four Tornado F.3 air defense fighters and twelve Jaguar GR.1A strike fighters—arrived in Dhahran, Saudi Arabia, and Thumrait, Oman, respectively, on 11 August (Tornado GR.1s were held back until 27 August, while Whitehall mandarins debated deploying offensive or defensive systems).[4] President

Mitterrand ordered French forces to the Gulf, as did Canada and Australia. Morocco, Syria, and Egypt also pledged troops, the first non–Gulf Cooperation Council Arab states to do so.[5] Even so, by mid-August diplomatic strictures were complicating and even delaying force deployment. The magnitude of the crisis and the unity of regional partners and allies helped greatly, however, with Austria, Bahrain, Egypt, Greece, India, Oman, Qatar, Saudi Arabia, Switzerland, Thailand, and the United Arab Emirates (UAE), joined subsequently by several eastern European nations, streamlining approval procedures or granting outright blanket authorization. Though some hindrances remained, the subsequent deployment generally went smoothly.[6] Bedding down incoming forces raised numerous issues, but most were resolved reasonably quickly. Furnishing air base security posed one challenge but was resolved through cooperation between arriving forces and host-nation personnel. To compensate for the shortage of foreign personnel that could be assigned to base-protection duties and to furnish needed counterintelligence expertise, Saudi Arabia and the various Gulf states assigned their own security forces—who were intimately familiar with local threats, something incoming foreign personnel were generally not—to work with American and other coalition forces.[7]

Coalition participation and contributions were crucial to combat success, from initial force deployments in August 1990 through the seven-week air–sea–land war itself. In addition to the United States, Saudi Arabia, and Great Britain, thirty-four other coalition nations and five more contributed soldiers, sailors, airmen, marines, special forces, medical personnel, ships, aircraft, tanks, artillery, mechanized vehicles, and other forms of aid or relieved the military burdens of other nations that made such contributions by deploying their own forces to critical regions (such as Belgian, Dutch, German, and Italian fighter deployments to Turkey). Forces fell under either the Saudi-led Joint Force Command (led by His Royal Highness General Khaled bin Sultan, Royal Saudi Air Defence Force) or US Central Command (CENTCOM, led by General H. Norman Schwarzkopf, US Army).

The coalition's first task was securing the safety of Saudi Arabia. Recognizing the urgency of assistance, Khaled and other senior officials drew up five requirements that coalition partners had to meet. They must

- be capable of high mobility and a speedy response;
- have a deterrent capability, even before reaching the kingdom;

- be able to protect supply lines to the kingdom, especially maritime routes;
- be strong enough to defend the kingdom and confront any possible attack;
- be able to contemplate quickly launching an offensive to free Kuwait.

"The inescapable conclusion," Khaled wrote after the war, "was that only the forces of the United States could meet these conditions."[8] But even the United States would require coalition assistance. There were also concerns related to Saudi autonomy and its uniquely Islamic culture—in particular the relationship among Saudi Arabia, the United States, and incoming coalition partners—as well as to the role of Saudi and American/coalition military commanders. On 9 August, King Fahd appointed Khaled as Joint Force commander, a coalition military leader equivalent to America's Schwarzkopf. Although Khaled recognized it was "inevitable that the command structure of the air campaign should reflect American supremacy," he "wanted to be sure that [he] knew 100-percent of what was going on."[9] Saudi leaders rigorously guarded their prerogatives of state power and freedom of action, insisting that any proposed military action be reviewed and approved first by the Saudi government and that once the crisis passed, the coalition members quickly remove their forces from the country.

"If the truth be told, the task we faced during the crisis was not winning the war against Saddam," Khaled wrote after the war. "In my view the greatest challenge we faced—and our failure was what Saddam was counting on— was to make sure that the members of the coalition worked together without friction or dispute."[10] "To their credit," British ambassador to Saudi Arabia Sir Alan Munro wrote, "both commanders [Khaled and Schwarzkopf] kept their sentiments towards each other within bounds."[11]

Upon first meeting Khaled, Schwarzkopf sensed unease beneath the pleasantries. "[Lieutenant General Charles "Chuck"] Horner [US Air Force (USAF)] and [Lieutenant General John] Yeosock [US Army] had guessed he [Khaled] was worried that with my higher rank and broad experience as a commander I'd try to bulldoze him or order him around," Schwarzkopf recalled. "I had no such intention, and, as Khaled recognized this in the days that followed, we became true friends and effective colleagues."[12] As a young cadet, Khaled had graduated from officer training at the Royal Military

Academy Sandhurst and then during his career had attended and graduated from the US Army's Command and General Staff College, the USAF's Air War College, and the US Navy's (USN) Naval Postgraduate School, experiences that gave him broad familiarity with both the British and American military services and systems. Thus, both by inclination and by background he was desirous of helping. Schwarzkopf recognized that Khaled, as a senior military officer and a member of the royal family, was uniquely positioned to break loose funding and support for the bedding down of incoming coalition forces. Khaled did subsequently help in many ways. At Horner's suggestion, he secured approval to colocate CENTCOM's forward headquarters within the Saudi Ministry of Defense and Aviation, thus placing Schwarzkopf's air, land, and maritime component commanders in close contact with their Saudi and Joint Force Command counterparts.[13] Working with Yeosock, Khaled formed a Coalition Coordination Communications and Integration Cell, linking the emergent Saudi Joint Force Command and US CENTCOM. The Saudis and CENTCOM also formed a Joint Directorate of Planning within the Ministry of Defense to develop combined plans.[14] Finally, CENTCOM and Saudi air defense personnel established a Combined Control and Reporting Center to cover Saudi Arabia's northeastern sector and to be staffed by American and Saudi personnel and liaison officers from other coalition partners.[15]

Britain and America's great historical experience in the Gulf in general and in Saudi Arabia in particular helped to smooth relations. The United States had a long heritage of supporting the Saudi military with facilities development, equipment, and training. A USAF study team had recommended formation of an independent Royal Saudi Air Force (RSAF) and provided both aircraft and a training mission for the construction of military facilities for the fledgling air force.[16] The Saudis turned to Britain for aircraft for a while following an initial denial of American F-5s and F-105s, but from President Richard Nixon's era onward, the United States sold the Saudis tanks, surface-to-air missiles, and aircraft through its Foreign Military Sales program.[17] For nearly a decade, American and Saudi commanders and crews had worked and flown together on their respective E-3s, furnishing early-warning coverage for Saudi Arabia's Eastern Province. When in 1985 legislators in the US Congress refused to sell the RSAF advanced F-15E attack aircraft, the Saudis turned to Britain once again, executing the Al-Yamamah arms deal that resulted in the RSAF acquiring Hawk trainers and Tornado strike aircraft and air defense fighters. Buying American and British arms meant, of course, accepting numbers of both nations' support and contractor personnel "in country," which again furnished a means of nurturing increasingly close bonds among the personnel and military services of all three nations.[18]

Khaled himself had trained and traveled extensively in Britain and America, and his specialty as an air defense officer well suited him to understanding the air issues of the campaign. So had many others. For example, Saudi brigadier general Prince Turki bin Nasser, an RSAF F-15C pilot commanding King Abd al-Aziz Air Base at Dhahran, had graduated from the USAF Air University. Not surprisingly, Nasser formed a strong partnership with Brigadier General John McBroom, commander of the USAF 1st Fighter Wing, whose F-15Cs deployed to Dhahran. "They solved a thousand problems every day," Horner recalled. "Frequently one would see USAF and RSAF repair teams helping one another, even if it meant that the two sergeants repairing the jet were a bearded Saudi and a fresh-faced American woman."[19]

Though their respective nations and services exercised command over coalition aircraft and airmen, Horner, as Schwarzkopf's Joint Force Air Component commander (JFACC), exercised control over them.[20] As Schwarzkopf put it bluntly to Horner's chief of plans, Brigadier General Buster Glosson, USAF, "If you aren't part of the air campaign under Horner, you don't fly."[21] Although this arrangement might have posed a serious conflict, coalition partners accepted it, and some, indeed, eagerly participated in it. Glosson recalled how Brigadier General Ahmad al-Sudairy, the chief of the RSAF's Air Operations Directorate, persistently asked him, "Is the RSAF carrying its share of the load?"[22] Indeed, al-Sudairy repeatedly pressed Glosson and Horner to assign the RSAF's Tornado force against very high-threat targets. When Glosson demurred, Horner told him, "You need to let al-Sudairy make the decision for the Saudis, not you."[23] In one case, the Saudi airman pointed out that since one high-threat target was near a mosque, it would be better if a Saudi aircraft carried out the mission than a Western one, lest the mosque suffer blast damage from any stray bomb.[24] After the war, reflecting on the Gulf's command-and-control issues, Air Chief Marshal Sir Patrick "Paddy" Hine, joint commander of all British forces in the Gulf War, concluded, "I know of no major disagreements from any of the other members of the Coalition with the way that the Americans put the air campaign and the daily task order together."[25] Indeed, although relations at the senior command level occasionally reflected parochial service cultures and sometimes divergent policy and political differences of the various coalition members, at the most important level—the airman-to-airman level—relations worked very well. "It did not matter if they were the Americans, British, French, Saudis, Kuwaitis, or from the United Arab Emirates," Horner recalled, adding:

> When everyone sat around the table at night, when we talked about what we were going to do and what needed to be done, every voice was equal and treated equally. . . . We could sit there and discuss things as airmen. We could talk about what made sense from an operational standpoint. The French commander, who had his Jaguars there with x-payload, x-range, and x-capability, would say, "Okay, we will take these targets here, here, and here." The Brits with their Tornados would say, "Okay, we'll take these." So we worked it just the way one would work it with a single force. . . . Our problems were not among the airmen; we never had a single problem among the airmen.[26]

On 27 August 1990, Horner stated that as JFACC he would "conduct, in the near term, a theatre air campaign to seize the initiative by attacking, isolating, and incapacitating the Iraqi military leadership and destroying Iraq's ability to conduct military operations."[27] Horner concentrated his planning staff in a special secure area in the basement of RSAF headquarters in Riyadh, nicknamed the "Black Hole" because, like its astronomical namesake, what took place within it could not escape. Headed by Brigadier General Glosson, this Black Hole staff included carefully chosen, highly experienced airman-planners from across the coalition. "Of course it was very easy to integrate the RAF guys and the Canadians into the planning cell, because they had all the clearances," Horner said after the war, adding, "but eventually we opened the doors to all members of the coalition: the Bahrainis, Saudis, Kuwaitis, everyone had full access and it paid off. Not only did we work well together to establish those links of trust, but everybody knew there was no hidden agenda."[28]

Horner rejected out-of-theater meddling, a legacy of his harrowing Vietnam experience during the ill-conceived and ineptly executed Operation Rolling Thunder (1965–1968), a veritable exemplar for airmen of "how not to" plan and conduct an air campaign. He likewise rejected a gradualist linear "start at the frontier, start at the beach" rollback campaign favored by the USN and US Marine Corps (USMC) incorporating service-specific, Vietnam-style "Route Packages" and other artificial zones of service exclusion and exclusivity.[29] Considering the objectives determined by Schwarzkopf as the theater CINC (commander in chief) and with inputs from Glosson's Black Hole, Horner produced a theater-wide, four-phase campaign against eleven target sets: Iraqi strategic air defense; chemical warfare and Scuds; leadership; Republican Guard; command, control, and communications; electricity; oil; railroads and bridges; airfields; ports; and military support, production, and storage. Coalition airmen, consistent with guidance from their own governments and their force's capabilities, flew against many of these sets of targets.[30]

In December 1990, Horner's airmen installed a hardened Tactical Air Control Center (TACC) in the basement of RSAF headquarters to oversee daily combat operations, including, as was often necessary, amending or changing the daily air tasking order that governed combat operations and, if necessary, adjusting for unforeseen or changing circumstances such as weather, revised intelligences, reactions to enemy action, and so on. In addition to Horner, Glosson, and other senior Americans, the TACC included other American commanders, action officers and support personnel from all American services, and representatives from various weapon-system communities, such as the F-117, B-52, F-15E, and so on. But it also included coalition commanders and key staff: Lieutenant General Ahmad al-Buhairi, RSAF; Air Vice Marshal William "Bill" Wratten, RAF; Major General Claude Solanet, Armée de l'air or French Air Force (FAF); Major General Mario Arpino, Aeronautica Militare Italiana; Lieutenant Colonel Abdullah al-Samdan (Kuwait); and Lieutenant Colonel John McNeil, Canadian Air Force (CAF).[31]

Although coalition airmen got along well, there were politically laden differences over rules of engagement (ROE) between British and American forces. Over Saudi Arabia, RAF Tornado F.3s flew combat air patrol for USAF E-3 Sentry airborne warning and control airplanes and RC-135 Rivet Joint signals intelligence collectors. British ROE required a dual confirmation on any inbound potentially hostile aircraft approaching these "high-value assets," whereas American ROE mandated only one confirmation. Some senior British civil and military officials, worrying over what they regarded (in Munro's words) as an innately American "cowboy approach," feared any reflexive shoot-down might "provoke a wider conflict."[32] The conflict quickly reached senior level: "If your boys can't take on hostile-approach aircraft, I really don't want them up there," Horner told General Sir Peter de la Billière, British Army, adding: "We'll deal with things ourselves because you'll be too late."[33]

De la Billière—who greatly admired Horner—fully appreciated the need for more realistic ROE. So, too, did Air Vice Marshal Wratten, dual-hatted as de la Billière's deputy commander and air commander in the Gulf, who quite rightly recalled years later that "the [high-value asset] itself could well identify a hostile aircraft."[34] But, to his frustration, de la Billière found that back in Whitehall "ministers seemed astonishingly reluctant to face reality or to understand the predicament we were in."[35] At the end of 1990, less than three weeks from the onset of hostilities, de la Billière wrote: "The level of ministerial indecision and looking backwards is appalling and desperately time-wasting. There is every likelihood that we shall stay behind while the Americans go to war and our ministers dither over their decisions."[36] Fortunately, thanks to Air Chief Marshal Hine back at High Wycombe in England and to Wratten in Riyadh, the ROE were changed.

Given the Iraqi regime's air and missile strength, coalition air planners knew well the first and most important task would be to gain control of the air through strikes targeting Iraq's integrated air defense network. Table 9.1 details coalition air power by nation, mission areas, and numbers; these data do not include US or North Atlantic Treaty Organization (NATO) forces outside theater, such as tankers, airlift, and B-52s in the United States, Europe, and the British Indian Ocean island Diego Garcia, which participated in Desert Shield's airlift and Desert Storm's airlift and strike operations.

Of the non-US coalition fighter and strike aircraft, 454 were combat available at the onset of the war, constituting a quarter of all coalition fighter and attack aircraft. Given the wide diversity of aircraft types, logistics support was challenging, and planners had to "deconflict" coalition Dassault Mirage F-1E operations in the first weeks to avoid possible "blue on blue" incidents given that the F-1E was one of the Iraqi Air Force's fighter mainstays.

From 7 August 1990 through mid-January 1991, American and coalition partner airmen flew a total of 132,029 sorties: 28,823 by partner air forces and 103,206 by American airmen; almost half of all coalition sorties flown during Desert Shield were for training.[37] Fully 30,423 of total coalition sorties were non-US-flown combat air patrol, air-to-ground, air-to-air, air-refueling, Airborne Warning and Control System, tactical/strategic reconnaissance,

Table 9.1.   Coalition Combat Air Forces by Mission Area, Gulf War, 1991

Nation	Type of Aircraft				
	Fighter/ Attack	Tanker	Airlift	Other	National Total
United States	1,323	285	175	207	1,990
Saudi Arabia	276	15	38	10	339
United Kingdom	57	9	3	8	77
France	44	3	12	7	66
Kuwait	40	—	3	—	43
Canada	24	1	12	2	39
Bahrain	24	—	—	—	24
Qatar	20	—	—	—	20
UAE	20	—	—	—	20
Italy	8	—	3	1	12
South Korea	—	—	5	—	5
New Zealand	—	—	2	—	2
Argentina	—	—	3	—	3
Totals	1,836	313	256	235	2,640

Sources: Data aggregated from Sir Patrick Hine, "Air Operations in the Gulf War," in *The War in the Air 1914–1994,* ed. Alan Stephens (Canberra: RAAF Air Power Studies Centre, 1994), 345–46, tables 1 and 2, and Diane Putney, *Airpower Advantage: Planning the Gulf War Air Campaign, 1989–1991* (Washington, DC: Air Force History and Museums Program, 2004), 306.

and airlift sorties by or for RSAF, RAF, CAF, FAF, and Aeronautica Militare Italiana airmen.[38] Earlier exposure to NATO, Central Command Air Forces, and various other training exercises, such as the USAF's Red Flag program at Nellis Air Force Base and the US Navy's Strike U program at Naval Air Station Fallon, greatly benefitted coalition synergy. British and Canadian airmen had flown in the Red Flag exercise in 1990, in which participants flew against threats simulating Soviet fighters and surface-to-air missiles, and British, Canadian, French, and Italian airmen had flown in the Red Flag exercise in 1989. Reflecting on Desert Storm, Wratten recalled, "All of the US forces and many of the coalition air forces had been through the Flag programs. We all spoke the same language. We knew the terminology. We were accustomed to force package thinking, and that is why it glued together remarkably well and extremely quickly."[39]

With the approval of Saudi authorities, B-52 crews flew training missions through Saudi airspace starting in August 1990.[40] USN airmen flew against UAE Dassault Mirage 2000 and Qatari Mirage F-1E fighters in dissimilar air combat training, the latter good preparation given the Iraqi air force's extensive F-1E inventory.[41] The USMC undertook extensive exercising of deployed aircraft, helicopter, and air–ground liaison forces in theater.[42] Difficult weather and blowing sand caused numerous problems for special-operations helicopter crews when they practiced nighttime, very-low-level operations. RAF Tornado crews discovered that their terrain-following radar could not always distinguish sand dunes from the hard desert floor. In other cases, pilots flew too aggressively at low altitude, leading to the loss of an American RF-4C and F-15E as well as of a British Jaguar later; in response, Horner raised the minimum fighter training altitude to 1,000 feet above ground level and called a mandatory "let it all hang out" safety meeting in Riyadh for wing commanders to raise issues, highlight problems, and identify solutions. After that, safety improved markedly.[43]

DESERT STORM: COALITION AIRMEN AT WAR

Following final—and futile—diplomatic attempts to resolve the political crisis with Iraq, the coalition went to war on 17 January 1991, with H-hour set at 3:00 a.m. local, the time when planners believed that Iraqi defenders would be least prepared.[44] Horner had kept the details of the air campaign plan very close, analysts noting that, "except for a handful of British and Saudi planners in the Black Hole, Coalition [sic] air forces remained in the dark about their targets until forty-eight hours or less before D-Day."[45]

The coalition partners' air power projection capabilities fell into three national groupings. The first consisted of those nations furnishing the fullest set of balanced capabilities, who also (and not surprisingly) were the most active of coalition air forces: USAF, USN, USMC, RSAF, and RAF. Together, the United States, the United Kingdom, and Saudi Arabia furnished more than 2,400 aircraft (representing 91 percent of the coalition total), flew 113,639 sorties (more than 95 percent of total coalition sorties), and sustained thirty-six losses (almost 95 percent of all coalition air losses).

The RSAF was the most active non-US coalition air force, flying 6,852 sorties—not quite 40 percent of all non-US coalition sorties. Excluding American forces, Saudi aircrews flew 36 percent of interdiction sorties,

Two USAF McDonnell-Douglas F-15C Eagles of the 33rd Tactical Fighter Wing flying in formation with an RSAF Northrop F-5E Tiger II fighter during Operation
Desert Storm. US Department of Defense photograph, at https://catalog.archives
.gov/id/6473253.

46 percent of counterair/air superiority sorties, 42 percent of airlift sorties,
31 percent of refueling sorties, and 35 percent of reconnaissance sorties.[46]
Most strike sorties were in southern Iraq and Kuwait, though Saudi Tornados
joined their British counterparts in attacking Iraqi airfields.[47] Saudi pilots
were active from the earliest days of the Gulf crisis, one Tornado pilot recalling later that "they were very, very good."[48] How good was revealed on
24 January when two Iraqi Mirage F-1s threatening coalition naval forces fell
to a Saudi F-15, likely preventing a *Stark*-like tragedy (in which the USS *Stark*
was almost sunk by two Mirage-launched Exocets during the Iran-Iraq War
in 1987, with thirty-seven killed and twenty-one injured) or worse.[49] Horner
stated bluntly after the war that the RSAF "performed magnificently."[50] As
one Saudi study concluded, "The RSAF entered the Gulf War as the most
modern air force in the region and acquitted itself well. It not only played a
crucial support role as the host nation's air force, but it also had the force

structure to play a role second only to the United States in command and control, reconnaissance and airlift and tactical air, and third behind Britain in air refueling."[51]

Next most active was the RAF, which—though it had extensive experience in the Gulf and Arabian Peninsula from 1945 on with air-supported special operations in small conflicts—was fighting its first major combat action since the Falklands War in 1982 and its first large-scale action since the Second World War.[52] Lieutenant General David A. Deptula, USAF, recalled that "from early on, the British were just extraordinarily cooperative and very, very helpful."[53] The RAF introduced some complementary capabilities the USAF did not possess: the JP233 runway cratering and airfield denial munition; the Air-Launched Antiradiation Missile; and a time-delay fuse, which, joined to UK-supplied 1,000-pound bombs, enabled USAF B-52s to undertake night airfield attacks with much greater effectiveness.[54]

In turn, the RAF benefitted from some of the weapons and avionics capabilities of American and other services. RAF Tornados received Identification Friend or Foe transponders and frequency-agile (hence jam-resistant) Have Quick radios, and they gained hand-held Global Positioning System satellite navigation receivers. The RAF also acquired US CBU-87 cluster bombs, which could be dropped from medium altitude, unlike the low-level British BL-755. Canada furnished the Bristol Aerospace (Canada) CRV-7 2.75-inch rocket, which was fired from pods carrying nineteen rockets, was extremely fast, and thus had a commendably flat trajectory, making it both highly accurate and an excellent antishipping weapon.[55]

The Panavia Tornado GR.1 was the RAF's principal strike aircraft, flying more than 1,500 offensive sorties (and 1,644 overall), dispensing 106 JP233s (submunition delivery systems), 3,641 unguided bombs, and 1,079 laser-guided bombs (LGBs), but at the cost of seven aircraft lost (one apparently to a premature bomb detonation), five crewmen killed, and another seven taken prisoner, a loss rate of not quite five aircraft per thousand sorties, higher than the loss rate of USAF F-105s over North Vietnam in 1967 at the peak of the infamous Rolling Thunder campaign.[56] The Tornados began their war by executing night low-level (180-foot), 500-kiloton counterairfield strikes, overflying runways, expending two JP233s per aircraft, and dispensing a total of 106 JP233s in 53 sorties executed over the first four days.

Crews flew through flak so thick that one Tornado pilot described it as "a wall-to-wall incandescent white curtain of light."[57] "No one else even approached the RAF in terms of courage," Horner recalled after the war. "Those guys were going through hell. I tried subtly to get them to back off and I could not do it. They kept saying, 'Give us the tough ones, give us the tough ones, give us the tough ones!'"[58]

Low-altitude losses were more than an RAF issue; the coalition flew most sorties at medium altitude, but two-thirds of losses were at low altitude.[59] On 23 January, following a coalition-wide tactics review directed by Horner and headed by Wratten (whom Horner considered "an absolute master of his profession" and Glosson regarded as "a super guy, very bright, absolutely top drawer"), Tornados (like other coalition strikers) switched to radar bombing from medium altitude, attacking radar sites, dumps and depots, and power stations and hunting Scuds, though with greatly reduced accuracy and, therefore, rising concern.[60] Ironically, initial planning had intended for the RAF's Tornados to drop LGBs, using "buddy lasing" from Buccaneer S.2Bs equipped with the Vietnam-legacy Pave Spike laser designator. But so crowded were Gulf airfields that Horner rejected bringing in the Buccaneer, agreeing instead to furnish USAF "buddy lasing" F-15Es. Then came the first Scud firings, triggering a "Great Scud Hunt" that took away any spare F-15Es, forcing deployment of the Buccaneer, hastily ordered to the Gulf.[61]

On 2 February, the Tornado force at last began dropping LGBs using buddy lasing from Buccaneers.[62] A week later technicians introduced two new GEC Ferranti Thermal Imaging Airborne Laser Designator (TIALD) pods into service. Unlike the day-only Pave Spike, TIALD worked at night as well. TIALD's first strike came on 10 February, and just two days later Air Vice Marshal (later Air Marshal Sir) Ian Macfadyen hailed it as "one of the success stories of this campaign," noting it was "working extremely well."[63] By war's end, TIALD-cued Tornados had scored 229 hits in 95 sorties (slightly more than one-third flown at night and with only one sortie aborted), shattering hardened aircraft shelters, cratering runways, and bombing ammunition, fuel-storage, and airfield facilities. Use of TIALD freed the twelve-plane Buccaneer force (which had flown 214 buddy-lasing sorties for the Tornados) to undertake thereafter its own self-designating bombing strikes. Together, the Tornado and Buccaneer introduced the RAF into the precision-weapon era.[64]

Although RAF Tornado operations gained the greatest media attention, the small Jaguar force flew 600 sorties against fielded Iraqi forces in the Kuwaiti theater of operations.[65] On 29 January, a flight of Jaguars returning from an attack on an Iraqi Silkworm antishipping missile site on the Kuwaiti coast spotted Iraqi patrol boats speeding south. In concert with Royal Navy Lynx helicopters (armed with Sea Skua antishipping missiles) cued by RAF Nimrod maritime patrol airplanes, they attacked the formation, sinking several boats. The strike opened the Bubiyan Turkey Shoot, a two-day air–naval action over which RAF, Royal Navy, CAF, and USN airmen sank or damaged more than two dozen Iraqi vessels, effectively destroying Saddam's navy.[66]

At sea, the RAF's Nimrod force searched for vessels attempting to run the UN blockade on Iraq during Desert Shield and then during Desert Storm surveyed Gulf waters for any vessels that might endanger the coalition, earning praise from the USN for having "contributed directly to the destruction of the Iraqi Navy."[67] The RAF's tanker force refueled many coalition aircraft, while its tanker transports, Hercules airlifters, and Chinook and Puma helicopters undertook a wide range of airlift and air support missions. As Group Captain Andrew Vallance wrote in the war's aftermath, "The Gulf War was first and foremost an air power war, and the RAF's contribution to the Allied air effort was significant and distinguished. The Service can take just pride in a remarkable feat of arms and a splendid professional achievement."[68]

The second national grouping consisted of nations that brought significant combat capabilities but were smaller in size or whose operations were more constrained because of the ROE imposed by their national governments. This group included France, Canada, and Italy. Together they furnished more than 100 aircraft (representing more than 4 percent of the coalition total), flew 3,797 sorties (more than 3 percent of total coalition sorties), and sustained one loss (an Aeronautica Militare Italiana Tornado IDS, its loss constituting almost 3 percent of total coalition air losses).

France sent the third-largest air expeditionary force to the Gulf, ranking behind the United States and Britain. Saudi and coalition relations with France were a constant challenge, however. "There was hardly an aspect of the French expeditionary force," Khaled wrote afterward, "which did not become a subject of argument."[69] French defense minister Jean-Pierre Chevènement, a founder of the Iraqi-French Friendship Society (and regarded by Khaled as "ideologically sympathetic to Saddam"), opposed intervention and obsessed over asserting French independence.[70] Lieutenant General de la Billière recalled: "Chevènement insisted that all command decisions must go through him in Paris—a requirement which made it impossible for the French to work closely with the Americans and condemned them to isolation."[71] Horner recalled a Saudi–American meeting with a French military delegation on

15 August during which "there [was] lots of cagey diplomatic talk": Horner resisted the temptation to say that "if they [the French] had been better allies in NATO, then we would not be having these 'getting to know you' sessions in Riyadh," and he instead gave the delegation "assurances that their national sovereignty will be respected, and as a sovereign nation they will be equal partners," leaving unspoken the condition that the French "will have to work with the Americans if they want in the game," something that French airmen, if not their political leaders, already realized.[72]

Although French forces deployed to the Gulf region beginning in September 1990, Paris did not commit to joining the coalition counteroffensive against Iraq until shortly before Desert Storm. "On instructions from Paris, [French commander Lieutenant General Michel] Roquejeoffre [French Army] wanted to place strict limits on French participation," Khaled recalled. "He wanted us to draw up and sign a sort of contract, defining the mission of his force within strict space/time parameters," adding, "It looked as if the French wanted to retain the right not to fight!"[73] Though "I respected and counted him as a friend," Schwarzkopf recalled, Roquejeoffre left Schwarzkopf wondering "whether we could count on the French to fight."[74] To Horner, who was trying to build a unified coalition air campaign, Roquejeoffre "was just bound and determined [that French airmen] could only fly over the French army."[75]

But events worked against Chevènement. In September, in an astonishing blunder Iraqi troops entered the residence of the French ambassador to Kuwait, seizing several persons, including the embassy's defense attaché. At a senior US-French meeting in Munich on 18 September, President Mitterrand bluntly informed Chevènement that if war came, French forces must fight under Schwarzkopf's command. On 16 January, the French Parliament approved by a vote of 523–43 a government resolution supporting military action by French forces under American command "for specific times and missions," mirroring French public opinion, which over the previous weeks had shifted from antiwar to prowar, a poll showing two-thirds of those interviewed favored military action.[76] When the coalition went to war, Chevènement was gone, replaced by Pierre Joxe, who ordered Roquejeoffre to work with Schwarzkopf. "With characteristic decisiveness, Norman immediately brought the French in on his plans and found them a real role in the Coalition line-up," de la Billière recalled, "so that instead of being very much on the margin, they had a genuine role in Operation Desert Storm."[77]

Chevènement had insisted French airmen could not colocate with the Americans and British, so Khaled sent them to Al Ahsa, a small civil airport that the Saudis then improved at their own expense. Commanded by Colonel (later General) Marc Amberg, Al Ahsa became the hub of Gallic air power projection.[78] By the outbreak of hostilities, the FAF had deployed more than 60 aircraft and helicopters, together with 1,300 airmen, the force being commanded by Major General Claude Solanet, reporting to Roquejeoffre.[79]

Though French airmen trained extensively with their coalition partners and offered their experiences and insights on Iraqi abilities and tactics, they were not "fully integrated into coalition planning and execution" until after Paris decided to commit to the counteroffensive.[80] "Dealing with the French was really, really interesting," Deptula recalled, noting one case where after he had designated a particular Iraqi target for French attack, an FAF colonel promptly informed him that "French politicians had refused any attacks on Iraq, but Kuwait was OK."[81] Given the "huge panoply of resources available," Deptula fortunately had other forces he could draw upon, though, he noted, the superb French-made Mach 1.75 AS-30L laser-homing missile, with a 528-pound warhead that could penetrate 6.5 feet of reinforced concrete at a range of 6.2 miles (10 kilometers) would have been very useful against some Iraqi targets.[82]

The first French air strike in the Gulf War began at 5:30 a.m. local time on 17 January 1991, when twelve Jaguars departed Al Ahsa to raid Kuwait's Ahmed Al Jaber airfield. They executed a spirited low-level attack, suffering four Jaguars damaged, including one nearly destroyed by an infrared-homing missile (likely an SA-7 Strela) and another penetrated by a round that hit the pilot's helmet, fortunately only dazing him.[83] After that, the FAF immediately went to medium altitude, and the TACC in Riyadh typically tasked the FAF with two air interdiction missions per day. The FAF's Jaguars and Mirage F-1CRs attacked across Kuwait and later in Iraq, targeting Republican Guard divisions, road and rail bridges crossing the Euphrates, and later air bases, including Tallil. The FAF flew its last combat sorties against Iraqi forces on 27 February 1991, having accomplished 2,258 sorties, including 531 interdiction, 570 counterair, 855 airlift, and 223 refueling ones.[84]

In retrospect, although the coalition undoubtedly benefitted from France's air–land–sea contribution, the problems caused by the French government's—in particular Jean-Pierre Chevènement's—obsessive desires for an independent role generated both distraction and discord. At the political level, French participation helped secure a broader unified front against Saddam's aggression. One recent study concludes, however, that at the military level "political reluctance to subordinate French forces to the joint force commander during the force build-up led to a military marginalisation of the FAF contingent."[85] But even under these conditions, strong in-theater airman-to-airman bonds generated mutual cooperation. "We just put them [French airmen] on the schedule and they flew where they were needed most and could be most effective," Horner recalled shortly after the war, the French airmen fighting with skill and élan—the mix of gallantry and verve traditionally associated with the Armée de l'air.[86]

Canada, like France, took on a limited role in the war and, again like France (though not so strongly), experienced dissension among its top leadership. On 10 August, Prime Minister Brian Mulroney announced that Canada would send a small, three-vessel naval force comprising two aging destroyers and a supply ship "to deter further Iraqi aggression."[87] The dispatch of its ships, named Operation Friction, initiated Canada's involvement in the Gulf. What initially was largely a symbolic and token naval presence, however, transformed significantly with the Mulroney government's decision a month later to dispatch a force of eighteen—later increased to twenty-four—CF-18 fighter-bombers (Canada's variant of the USN-USMC F/A-18) to protect the ships, under Operation Scimitar.[88] Designated the Canadian Air Task Group–Middle East, the force drew primarily upon the German-based 1st Canadian Air Division, then commanded by CAF Brigadier General Jean Boyle. Though many units contributed to this Canadian air force over its short existence, its core on the eve of Desert Storm blended 416 Lynx Squadron and 439 Tiger Squadron, inspiring the name "Desert Cats."[89] The first CF-18s deployed from Germany on 6 October via RAF Akrotiri in Cyprus, landing at Doha on 7 October, undertaking their first local-familiarization sorties the next day, and flying their first protective combat air patrols on 9 October.[90]

Like the other coalition partners, the Canadians trained aggressively and hard, flying against American and other coalition aircraft in dissimilar air combat exercises, fighter sweep and escort tactics, and air-to-ground attack (not then considered a likely mission given Canadian policy). Canadian weapons controllers also served in E-3s of the USAF's 552nd Airborne Warning and Control System wing (indeed, the Canadian crewmen in this wing were airborne in E-3s during thirty-five of the coalition's shoot-downs of

Iraqi aircraft).[91] As noted previously, Canadian airmen served in the TACC, being perfectly familiar with the air tasking order process and product.

Administratively, the Canadian Air Task Group–Middle East reported to Commodore Kenneth J. Summers, a naval officer who assumed command of Canadian Forces Middle East headquarters on 6 November 1990. The Desert Cats began flying patrols over the northern and central Persian Gulf. On 14 October, a week after having arrived in theater, Desert Cat CF-18s scared off two Iraqi aircraft, earning a heartfelt "Way to go, Canada!" from the USN, which had been sensitized to the Iraqi antishipping threat since the near sinking of the USS *Stark* four years earlier.[92]

The Canadians' now demonstrated ability to ensure fleet safety freed USMC F/A-18 pilots to concentrate on preparing for hostilities, which, of course, opened on 17 January.[93] But at that time the Mulroney government was still debating critics in the Canadian Parliament: it did not secure legislative approval for supporting offensive action by the coalition until 20 January.[94] Now truly at war, Canadian airmen sought a more expansive combat role, leading one frustrated CAF officer to appeal directly to Horner, complaining that Summers had "gone bonkers" in his insistence the Desert Cats must continue to restrict themselves simply to fleet defense.[95] Unhappy, Horner met with Summers (recalling afterward, "I had a rather difficult conversation, but fortunately I outranked him by two stars") and won some concessions for Desert Cat participation in sweep and escort missions, though none for air-to-ground missions, which was precisely what the coalition most needed.[96] So, dissatisfied with his meeting, Horner craftily arranged to sit beside Mary Collins, Canada's associate minister of national defence, then visiting the Gulf, at a diplomatic function. In response to her questions about the air campaign, he explained to her that the Desert Cats were effectively being sidetracked from playing a more substantial role in the larger war.[97] The message struck home, Horner recalling that "the next day the Canadian squadron was on the operational plan—dropping bombs."[98]

The Desert Cats fought tenaciously and well over the remainder of the Gulf War, though their core mission remained fleet defense.[99] Following the launching of the ground offensive on 24 February, the Desert Cats bombed artillery positions, troop concentrations, and road convoys, employing radar bombing from altitude to avoid Iraqi battlefield antiaircraft artillery and missiles. Like much of the coalition—including most American strike aircraft— CF-18s could not then drop precision weapons (though a USN offer to buddy lase for them so they could undertake LGB strikes was in the works at the war's end). Thus, dropping unguided dumb bombs by radar from altitude produced inconclusive results.[100] Overall, the CAF flew a total of 1,302 combat sorties, including 48 interdiction, 837 counterair, 277 airlift, and 64 refueling ones.[101] In retrospect, as with the Armée de l'air, the CAF might have been used more efficiently if fully incorporated within the coalition air forces rather than considered a shore-based appendage of the Canadian and US navies, a situation reflecting Horner's postwar judgment that "our problems were not among the airmen" but rather "between airmen and other services."[102]

Italy was the only other NATO nation aside from Britain and France to engage in offensive combat operations against Iraq. On 1 September 1990, the Italian government announced that it would deploy Tornado IDS strike fighters and support aircraft to the Gulf. The deployment, Operazione Locusta, began on September 25, undertaken by the Golfo Persico Reparto Volo Autonomo Aeronautica Militare (Persian Gulf Autonomous Air Force Flight Unit), a special expeditionary unit under General Mario Arpino. That day eight Tornados and crews departed Gioia del Colle for Al Dhafra. The Tornado detachment, commanded by Colonel Mario Redditi, set to work training for the war to come.[103]

By the opening night of the war, Italy's conflicted government still had not decided whether it would go to war, and so Arpino met with Horner. "My government can't make up its mind whether it's going to be in the war or not," he told the JFACC, adding morosely: "I can't fly until I get the go-ahead." Horner replied, "Mario, we all have governments; don't worry about it."[104] The Italian contingent accordingly flew on the second night of the war, after Rome had decided to commit. During the contingent's first strike against an arms depot a dozen miles north of Kuwait City, tanking difficulties prevented all but one aircraft from refueling, and that lone attacker was shot down by antiaircraft fire, its crew ejecting into captivity. Undeterred, the Locusta force continued to strike at Kuwaiti and Iraqi targets through the end of the war. Italy also sent a detachment of RF-104G reconnaissance fighters to Turkey, the springboard for Proven Force, a self-contained air expeditionary wing of USAF F-111Es and other aircraft operating against northern Iraq.[105]

Finally, the third grouping consisted of various Gulf-region air forces: Bahrain, Kuwait, Qatar, and UAE. Although they lacked the overall broad range of capabilities afforded by larger services, their contribution was still most valuable. Together they contributed more than 100 aircraft (representing more than 4 percent of the coalition total), flew 1,225 sorties (more than one percent of total coalition sorties), and sustained one loss (a Kuwaiti A-4).[106] The Kuwaiti air force is particularly deserving of mention, for it was essentially an "expeditionary" air arm, having been forced under hostile fire from its bases, during which it made a fighting withdrawal, a testament to the courage and gallantry of its airmen.

THE REFUELING CHALLENGE

Air refueling constituted both a key enabling requirement and a potential coalition weakness, for planners seriously underestimated the coalition's tanking needs.[107] Saudi Arabia, UAE, and Oman made a particularly substantial contribution to the coalition by providing 1.76 billion gallons of fuel for air, sea, and land operations, then worth approximately US$2.1 billion (US$4.1 billion in 2019). This fuel donation amounted to close to 94 percent of the total fuel consumed during Desert Shield and Desert Storm, which otherwise would have necessitated an extensive sealift of fuel from the United States, thus complicating transportation, delivery, and storage.[108] Despite the plentiful local supply of fuel, interoperability challenges remained because of differing fuel types, differing refueling systems (the boom-and-receptacle system or probe-and-drogue system), airspace deconfliction, and certification requirements for non-US airmen and aircraft to receive fuel from American tankers. As analysts noted later, "Coalition air warfare would have been better served if certification and training efforts of Allied receivers had been conducted in peacetime," recommending changes "to ensure interoperability in a future crisis."[109]

Different refueling systems posed a particular challenge. USAF KC-10 and KC-135 tankers had a "boomer" (boom operator) "flying" a fixed boom and plugging it into the refueling ports of strike aircraft that pilots had maneuvered into position. USN and USMC aircraft as well as many other coalition and Special Operations Command Central aircraft and helicopters had either fixed or retractable refueling probes that necessitated the pilot maneuver the aircraft to plug into a drogue "basket" receptacle streamed from a tanker with a hose and reel or from a USAF KC-135 or KC-10 with a drogue basket attached to its boom.[110] The danger—encountered on several occasions, though not, fortunately, resulting in any losses—existed that a low-fuel aircraft might approach a tanker that could not refuel it because of receptacle differences. For that reason, hose-and-drogue-basket tankers—the USMC and Saudi KC-130; the British VC-10K, TriStar K1, and Victor K2; the Canadian CC-137; the French KC-135FR (which employed a boom-attached drogue basket); and the USN KA-6D and S-3B—were crucial to coalition safety, efficiency, and operational success.

During Desert Storm, coalition tankers off-loaded almost 107 million gallons of fuel to more than 43,000 receivers, an average of approximately 2.5 million gallons passed in 1,000 refuelings per day during nearly 370 daily tanker sorties.[111] At times, up to six layers of tankers were refueling aircraft, their tracks separated vertically by as little as 1,000 feet.[112] (Just a single seven-B-52 cruise-missile launch mission from the United States to Iraq and back, for example, necessitated five refuelings that involved 57 tanker sorties.[113]) Generally, the USAF, USN, USMC, RAF, FAF, CAF, and RSAF refueled their own aircraft, but they also on occasion fueled their coalition partners' aircraft as needed. USAF KC-10 and KC-135 tankers refueled aircraft of Bahrain, Canada, Italy, Oman, Saudi Arabia, and UAE as well as of the USN and USMC. The RAF likewise primarily refueled its own Tornados, Jaguars, and Buccaneers but also refueled Canadian CF-18s, Saudi Tornados, and USN A-6s and F-14s.[114]

DESERT STORM BY THE NUMBERS

Through to the end of hostilities, the coalition flew 118,661 sorties—101,370 (85.43 percent) by US airmen and 17,291 (14.57 percent) by Saudi, British, French, Canadian, Kuwaiti, Bahraini, Italian, UAE, and Qatari airmen—an average of 2,697 sorties per day (see tables 9.2 and 9.3).[115] Non-US coalition airmen flew

- 14.57 percent of all combat sorties;
- 12.10 percent of all interdiction/battlefield air interdiction sorties;
- 24.54 percent of all counterair/air superiority (offensive counterair/defensive counterair and suppression of enemy air defenses, escort, combat air patrol, airfield attack, SEAD, etc.) sorties;
- 19.97 percent of all airlift sorties;
- 10.57 percent of all reconnaissance sorties;
- 9.89 percent of all refueling sorties;
- 2.74 percent of all electronic-warfare sorties; and
- 4.27 percent of all command, control, and communication sorties.

Table 9.2. US Sorties by Mission Area and Service, Gulf War, 1991

Service	Interdiction	CAS FAC	Counterair	Airlift	Reconnaissance	Refueling	Special Operations Forces	Electronic Warfare	C3	Misc.	Service Totals
Air Force	24,292	2,120	10,980	16,628	1,311	11,024	134	1,578	604	735	69,406
Navy	5,060	21	6,181	0	1,431	2,782	3	265	1,143	1,417	18,303
Marines	4,264	3,956	757	9	3	461	1	343	157	732	10,683
Army	0	0	0	201	147	0	0	568	0	0	916
SOCCENT*	32	31	0	19	2	56	808	84	0	230	1,262
CRAF*	0	0	0	800	0	0	0	0	0	0	800
US-Only Total	33,648	6,128	17,918	17,657	2,894	14,323	946	2,838	1,904	3,114	101,370
Coalition Total	38,277	6,128	23,745	22,064	3,236	15,895	948	2,918	1,989	3,461	118,661

* SOCCENT = Special Operations Command Central. CRAF = Civil Reserve Air Fleet.

Source: Data extracted and/or computed from *Gulf War Air Power Survey: Statistical Compendium and Chronology*, vol. 5 (Washington, DC: US Government Printing Office, 1993), tables 66–71, pp. 243–78. US airlift includes 800 CRAF sorties flown during the war. Army sorties are for fixed-wing assets only; the US Army records helicopter operations by flight hour, not by sortie.

Table 9.3. Coalition Sorties (Excluding US Sorties) by Mission Area and Nation

Coalition Nation	Interdiction	CAS FAC	Counterair	Airlift	Reconnaissance	Refueling	Special Operations Forces	Electronic Warfare	C3	Misc.	Nation Totals
Saudi Arabia	1,656	0	2,668	1,829	118	485	0	0	85	11	6,852
United Kingdom	1,256	0	1,586	1,384	156	711	0*	80	0	244	5,417
France	531	0	570	855	62	223	1	0	0	16	2,258
Canada	48	0	837	277	0	64	0	0	0	76	1,302
Kuwait	780	0	0	0	0	0	0	0	0	0	780
Bahrain	122	0	166	4	0	0	1	0	0	0	293
Italy	135	0	0	13	0	89	0	0	0	0	237
UAE	58	0	0	45	6	0	0	0	0	0	109
Qatar	43	0	0	0	0	0	0	0	0	0	43
Non-US Total	4,629	0	5,827	4,407	342	1,572	2	80	85	347	17,291
Coalition Total	38,277	6,128	23,745	22,064	3,236	15,895	948	2,918	1,989	3,461	118,661

Sources: Data extracted and/or computed from *Gulf War Air Power Survey: Statistical Compendium and Chronology*, vol. 5 (Washington, DC: US Government Printing Office, 1993), table 64, pp. 232–33. UK figure for special operations cannot be 0 because RAF Chinook helicopters supported Special Air Service counter-Scud insertions into western Iraq. See Andy McNab, *Bravo Two Zero* (London: Transworld, 1994), 81–92.

Through the length of the war, Iraq lost several hundred aircraft to offensive counterair bombing sorties. Another thirty-eight fell in air-to-air combat.[116] In turn, the coalition suffered thirty-eight combat losses. Including only combat-strike sorties into hostile fire, the coalition lost a little more than one aircraft for every 2,000 sorties flown. Besides those aircraft that were shot down or that disappeared over hostile territory from unknown causes, the coalition suffered forty-eight aircraft damaged by enemy fire.[117]

Conclusion

In retrospect, of two conclusions regarding the Gulf War there was then and is now no doubt. First, despite numerous studies and reports by military professionals, civilian analysts, journalists, and academics, nothing has emerged to refute President George H. W. Bush's pronouncement in his graduation address to the United States Air Force Academy Class of 1991 that "Gulf Lesson One is the value of air power."[118] Second, the war reinforced the enduring importance of coalition warfare. If certainly not perfect, the coalition worked; the partners came together to provide support, basing, unity, and combat strength. Although the war illustrated challenges to interoperability—communications, planning/targeting, tanking, logistics—none was insurmountable, and all forces involved—including those of the United States—adjusted their defense structures accordingly after the war. For all of that, for the lessons it offers in responding to aggressors, and for the affirmation it furnishes for coalition warfare, the Gulf War of 1991 remains deserving of study.

Finally, the Gulf War reminds us that long-simmering national and regional rivalries, often masked by more public squabbles among greater powers, can erupt with fearsome suddenness. In the fall of 1989, for example, Americans were debating how the nation would spend the "peace dividend" now that the Cold War was effectively over. Just a year later, more than 500,000 soldiers, sailors, airmen, and marines were deployed to a region few had considered for a high-tech war involving potential chemical, biological, and missile attacks, a war that took the fullest measure of American and coalition partner nonatomic power to confront, including stealth-attack aircraft; electronic combat; cruise missiles; space-based warning, cuing, communication, and navigation; and a global-ranging tanker and airlift bridge.

In the preparations for and execution of that war, national political and military leaders, planners, diplomats, and other decision makers—plus all those "at the pointy end"—had to confront not only the classic problems of coalition warfare but also issues too often neglected: the widely differing social, cultural, and religious perspectives of various partners that colored and influenced day-to-day operations and relations during Desert Shield and on into Desert Storm. In all these varied respects, Desert Shield and Desert Storm still have much to offer future campaign planners and those charged with delivering responsive, flexible, and overwhelming air and space power against a foe.

NOTES

1. Air Vice Marshal R. A. Mason, "The Air War in the Gulf," *Survival* 35, no. 3 (May–June 1991): 211.

2. Lawrence Freedman and Efraim Karsh, *The Gulf Conflict, 1990–1991: Diplomacy and War in the New World Order* (Princeton, NJ: Princeton University Press, 1993), 393–409; General H. Norman Schwarzkopf, with Peter Petre, *It Doesn't Take a Hero: The Autobiography of General H. Norman Schwarzkopf* (New York: Bantam Books, 1992), 452–53, 466–70; His Royal Highness General Khaled bin Sultan, with Patrick Seale, *Desert Warrior: A Personal View of the Gulf War by the Joint Forces Commander* (New York: Harper Collins, 1995), 403–12; General Sir Peter de la Billière, *Storm Command: A Personal Account of the Gulf War* (London: Harper Collins, 1992), 283–303; Tom Clancy and General Chuck Horner, *Every Man a Tiger: The Gulf War Air Campaign* (New York: Putnam's, 1999), 486–500; General Buster Glosson, *War with Iraq: Critical Lessons* (Charlotte, NC: Glosson Family Foundation, 2003), 260–83.

3. Donald B. Rice, "Air Force Restructure," in Office of the Secretary of the Air Force, *Toward the Future—Global Reach–Global Power: US Air Force White Papers, 1989–1992* (Washington, DC: Headquarters USAF, 1992), 63–71; Merrill A. McPeak, "Tomorrow's Air Force," in *Selected Works: 1990–1994* (Maxwell Air Force Base, AL: Air University Press, 1995), 67–113; Thomas A. Keaney and Eliot A. Cohen, *Gulf War Air Power Survey: Summary Report* (Washington, DC: US Government Printing Office [GPO], 1993), 7–8 (*Gulf War Air Power Survey* hereafter given as *GWAPS*); Sebastian Cox and Sebastian Ritchie, "The Gulf War and UK Air Power Doctrine and Practice," in *Air Power History: Turning Points from Kitty Hawk to Kosovo*, ed. Sebastian Cox and Peter Gray (London: Frank Cass, 2002), 287–300; Sebastian Ritchie, "The Royal Air Force and the First Gulf War, 1990–1991: A Case Study in the Identification and Implementation of Air Power Lessons," *Air Power Review* 17, no. 1 (Spring 2014): 36–53; Benjamin S. Lambeth, *Desert Storm and Its Meaning: The View from Moscow* (Santa Monica, CA: RAND, 1992), vii, 69–96; Roger Cliff, "Development of the PLAAF's Doctrine," in *The Chinese Air Force: Evolving Concepts, Roles, and Capabilities*, ed. Richard P. Hallion, Roger Cliff, and Phillip C. Saunders (Washington, DC: National Defense University Press, 2012), 153.

4. Cox and Ritchie, "The Gulf War and UK Air Power Doctrine and Practice," 292–93; "Britain and the Gulf War—Chronology," *Air Power Review* 19, no. 2 (Summer 2016): 13–14.

5. *GWAPS: Operations and Effects and Effectiveness,* vol. 2 (Washington, DC: US GPO, 1993), 16; Air Chief Marshal Sir Patrick Hine, "Air Operations in the Gulf War," in *The War in the Air: 1914–1994,* ed. Alan Stephens (Canberra: Air Power Studies Centre, 1994), 339–40; Edward J. Marolda and Robert J. Schneller Jr., *Shield and Sword: The United States Navy and the Persian Gulf War* (Washington, DC: Naval Historical Center, 1998), 61–62; Sir Alan Munro, *Arab Storm: Politics and Diplomacy behind the Gulf War* (London: I. B. Tauris, 2006), 58–59, 62–63; Freedman and Karsh, *The Gulf Conflict,* 110–15.

6. James K. Matthews and Cora J. Holt, *So Many, so Much, so Far, so Fast: The United States Transportation Command and Strategic Deployment for Operation Desert Shield/Desert Storm* (Washington, DC: US Department of Defense, 1996), 42–49, 51–57.

7. *GWAPS: Logistics and Support,* vol. 3 (Washington, DC: US GPO, 1993), 46–50, 234; Colonel Edward J. Hagerty, *The Air Force Office of Special Investigations, 1948–2000* (Washington, DC: USAF Office of Special Investigations, 2008), 358–61.

8. Khaled, *Desert Warrior,* 22.

9. Ibid., 325, 338.

10. Khaled, *Desert Warrior,* 264–65.

11. Munro, *Arab Storm,* 321.

12. Schwarzkopf, *It Doesn't Take a Hero,* 330.

13. Clancy and Horner, *Every Man a Tiger,* 195, 197–99.

14. Lieutenant Colonel Turki K. Al Saud, "The Royal Saudi Air Force and Long-Term Saudi National Defense: A Strategic Vision," master's thesis, US Marine Corps Command and Staff College, 2002, 19.

15. *GWAPS: A Statistical Compendium and Chronology,* vol. 5 (Washington, DC: US GPO, 1993), 136.

16. Al Saud, "The Royal Saudi Air Force," 3.

17. Ibid., 3–6; Major General Richard V. Secord, with Jay Wurts, *Honored and Betrayed: Irangate, Covert Affairs, and the Secret War in Laos* (New York: Wiley, 1992), 143–45, 166–74; David B. Ottaway, *The King's Messenger: Prince Bandar bin Sultan and America's Tangled Relationship with Saudi Arabia* (New York: Walker, 2008), 29–54; John E. Peterson, *Defending Arabia* (New York: St. Martin's Press, 1986), 87–114, 147–90.

18. Al Saud, "The Royal Saudi Air Force," 3–6; Clancy and Horner, *Every Man a Tiger,* 190.

19. Clancy and Horner, *Every Man a Tiger,* 206.

20. Rear Admiral James A. Winnefeld and Dana J. Johnson, *Joint Air Operations: Pursuit of Unity in Command and Control, 1942–1991* (Annapolis, MD: Naval Institute Press, 1993), 109.

21. *GWAPS: Operations and Effects and Effectiveness,* 39–40.

22. Glosson, *War with Iraq,* 138.

23. Ibid.

24. Ibid.

25. Comment by Air Chief Marshal Sir Patrick Hine, RAF, in response to Air Marshal Ray Funnell, RAAF, in Sir Patrick Hine, "Air Operations in the Gulf War," in *The War in the Air,* ed. Stephens, 357.

26. General Chuck Horner, "Address," in *Seeing Off the Bear: Anglo-American Air Power Cooperation during the Cold War,* ed. Roger G. Miller (Washington, DC: Air Force History and Museums Program, 1995), 145.

27. Quoted in *GWAPS: Operations and Effects and Effectiveness,* 40.

28. Horner, "Address," 147.

29. *GWAPS: Operations and Effects and Effectiveness,* 25, 39–40; Marolda and Schneller, *Shield and Storm,* 114–38; Clancy and Horner, *Every Man a Tiger,* 216–17.

30. Diane Putney, *Airpower Advantage: Planning the Gulf War Air Campaign, 1989–1991* (Washington, DC: Air Force History and Museums Program, 2004), 305–9; Glosson, *War with Iraq,* viii–ix; Lieutenant General David A. Deptula, USAF (ret.), conversation with the author, February 3, 2019; Major Jean Morin and Lieutenant Commander Richard H. Gimblett, *Operation Friction 1990–1991* (Toronto: Dundurn Press, 1997), 104–5; Marolda and Schneller, *Shield and Sword,* 146.

31. Clancy and Horner, *Every Man a Tiger,* 314–16; Glosson, *War with Iraq,* viii–ix; Deptula, conversation.

32. Munro, *Arab Storm,* 82–84; Charles Allen, *Thunder and Lightning: The RAF in the Gulf: Personal Experiences of War* (London: Warner Books, Her Majesty's Stationery Office, 1991), 33.

33. Quoted in de la Billière, *Storm Command,* 175.

34. Air Chief Marshal Sir William Wratten, RAF (ret.), statement in "Britain and the 1991 Gulf War Witness Seminar," *Air Power Review* 19, no. 2 (Summer 2016): 67.

35. De la Billière, *Storm Command,* 174–75.

36. Quoted in ibid., 177.

37. Calculated from *GWAPS: Statistical Compendium and Chronology,* table 46, p. 140.

38. Calculated from *GWAPS: Statistical Compendium and Chronology,* table 58, pp. 206–7.

39. Wratten, statement in "Britain and the 1991 Gulf War Witness Seminar," 67; *GWAPS: Weapons, Tactics, and Training and Space Operations,* vol. 4, part I (Washington, DC: GPO, 1993), appendix F, "Flag Exercises," 421–22.

40. *GWAPS: Weapons, Tactics, and Training,* appendix G, "B-52 Training," 437–40.

41. *GWAPS: Weapons, Tactics, and Training,* appendix I, "USN–USMC Training," 450.

42. *GWAPS: Weapons, Tactics, and Training,* appendix I, "USN–USMC Training," 452–53.

43. *GWAPS: Operations and Effects and Effectiveness,* 46; *GWAPS: Weapons, Tactics, and Training,* 350; Clancy and Horner, *Every Man a Tiger,* 295–97. The phrase "let it all hang out" is from Horner in Clancy and Horner, *Every Man a Tiger,* 296.

44. *GWAPS: Operations and Effects and Effectiveness,* 119.

45. Ibid., 48.

46. *GWAPS: Statistical Compendium and Chronology,* table 64, pp. 232–33.

47. Al Saud, "The Royal Saudi Air Force," 19; *GWAPS: Operations and Effects and Effectiveness,* 117.

48. Allen, *Thunder and Lightning,* 27.

49. *GWAPS: Operations and Effects and Effectiveness,* 334; *GWAPS: Weapons, Tactics, and Training,* 36, 201; *GWAPS: Statistical Compendium and Chronology,* 174; Marolda and Schneller, *Shield and Storm,* 205–8; Khaled, *Desert Warrior,* 358–59.

50. Clancy and Horner, *Every Man a Tiger,* 429; *GWAPS: Operations and Effects and Effectiveness,* 33 n. 62.

51. Al Saud, "The Royal Saudi Air Force," 18.

52. Sebastian Ritchie, *The RAF, Small Wars, and Insurgencies: Later Colonial Operations, 1945–1975* (Whitehall, UK: Air Historical Branch, Ministry of Defense, 2011), 45–58, 78–94, 120–33.

53. Deptula, conversation.

54. Ibid. See also *GWAPS: Weapons, Tactics, and Training,* 51, 83, 105; Cox and Ritchie, "The Gulf War and UK Air Power Doctrine and Practice," 296; Sir Patrick Hine, "Despatch," *London Gazette,* 28 June 1991; Group Captain Andrew Vallance, "Air Power in the Gulf War—the RAF Contribution," *Air Clues: The Royal Air Force Magazine* 45, no. 7 (July 1991): 251–54; "Britain and the Gulf War—Chronology," 17–19.

55. Major Frank J. Rossi, "The 'Capabilities Gap' in Desert Storm," master's thesis, School of Advanced Airpower Studies, 2001, 17–19.

56. Hine, "Despatch"; *GWAPS: Statistical Compendium and Chronology,* table 107, p. 345; *GWAPS: Weapons, Tactics, and Training,* 64; Vallance, "Air Power in the Gulf War," 253.

57. Allen, *Thunder and Lightning,* 78.

58. Horner, "Address," 149.

59. Clancy and Horner, *Every Man a Tiger,* 355; Vallance, "Air Power in the Gulf War," 251.

60. Horner, "Address," 149; Glosson, *War with Iraq,* 139. See also Clancy and Horner, *Every Man a Tiger,* 355, and Hine, "Despatch."

61. Wratten, statement in "Britain and the 1991 Gulf War Witness Seminar," 70.

62. Tornado mission report extract for As Samawah attack, 2 February 1991, *Air Power Review* 19, no. 2 (Summer 2016): 170–71.

63. Peter Binham, "TIALD—a Casebook of Cooperation," *The Hawk* 1992, 21.

64. Hine, "Despatch"; Binham, "TIALD," 28; Vallance, "Air Power in the Gulf War," 252; *GWAPS: Weapons, Tactics, and Training,* 46, 54, 57, 60, 63–65; Air Commodore Alistair Byford, "Operation *Granby* and the Dawn of Precision in the Royal Air Force: A Personal Perspective," *Air Power Review* 21, no. 2 (Summer 2018): 146–60.

65. *GWAPS: Weapons, Tactics, and Training,* 65; *GWAPS: Statistical Compendium and Chronology,* table 101, p. 341; Hine, "Despatch"; Vallance, "Air Power in the Gulf War," 253; Jaguar mission report extract for SA-2 site attack, 19 January 1991, *Air Power Review* 19, no. 2 (Summer 2016): 179; Dave Windle and Martin Bowman, *Sepecat Jaguar: Tactical Support and Maritime Strike Fighter* (Barnsley, UK: Pen & Sword Aviation, 2010), 60.

66. Hine, "Despatch"; Marolda and Schneller, *Shield and Sword,* 227–31; Vallance, "Air Power in the Gulf War," 253.

67. Sebastian Ritchie, "Operation *Granby:* Maritime Air Reconnaissance," *Air Power Review* 19, no. 2 (Summer 2016): 226–27.

68. Vallance, "Air Power in the Gulf War," 254.

69. Khaled, *Desert Warrior,* 266.

70. Ibid., 268.

71. De la Billière, *Storm Command,* 50–51.

72. Clancy and Horner, *Every Man a Tiger,* 228.

73. Khaled, *Desert Warrior,* 270–71.

74. Schwarzkopf, *It Doesn't Take a Hero,* 390.

75. Horner, "Address," 146.

76. Freedman and Karsh, *The Gulf Conflict,* 349–51.

77. De la Billière, *Storm Command,* 160.

78. Pierre Bayle, "Les aviateurs de Daguet," 26 February 2011, at https://www.site-daguet.fr/historique/opération-daguet-20-ans-après/les-aviateurs-de-daguet/.

79. Eric Micheletti, *Operation Daguet* (Hong Kong: Concord, 1991), 1–2.

80. *GWAPS: Statistical Compendium and Chronology,* 42.

81. Deptula, conversation.

82. Ibid.; Micheletti, *Operation Daguet,* 2, 22.

83. Micheletti, *Operation Daguet,* 2; Patrick Ehrhardt and Jean-Pierre Hoehn, "Guerre du Golfe: L'oeil du Daguet et la griffe du Jaguar," *Le Fana de l'aviation* 282 (May 1993): 14–26.

84. *GWAPS: Statistical Compendium and Chronology,* table 64, pp. 232–33. Numbers differ among sources, but I accept the numbers given in *GWAPS.*

85. Christian F. Anrig, *The Quest for Relevant Air Power: Continental European Responses to the Air Power Challenges of the Post–Cold War Era* (Maxwell Air Force Base, AL: Air University Press, 2011), 102.

86. Horner, "Address," 146.

87. Richard H. Gimblett, "Rethinking Maritime Air," in *Big Sky, Little Air Force,* ed. W. A. Marsh (Trenton: Canadian Forces Aerospace Warfare Centre, 2011), 125.

88. Freedman and Karsh, *The Gulf Conflict,* 478 n. 24.

89. Morin and Gimblett, *Operation Friction,* 97, 151–52.

90. David N. Deere, preface to *Desert Cats: The Canadian Fighter Squadron in the Gulf War,* ed. David N. Deere (Stoney Creek ON: Fortress, 1991), 10.

91. Morin and Gimblett, *Operation Friction,* 165.

92. Marolda and Schneller, *Shield and Sword,* 147.

93. Morin and Gimblett, *Operation Friction,* 160–61.

94. Freedman and Karsh, *The Gulf Conflict,* 349.

95. Horner, "Address," 145.

96. Ibid.

97. Ibid.

98. Ibid.; Morin and Gimblett, *Operation Friction,* 104–5, 114–15.

99. Morin and Gimblett, *Operation Friction,* 175.

100. Ibid., 160–78; Deere, *Desert Cats,* passim; *GWAPS: Statistical Compendium and Chronology,* table 64, pp. 232–33.

101. Lieutenant Colonel Samuel J. Walker, *Interoperability at the Speed of Sound: Canada–United States Aerospace Cooperation: Modernizing the CF-18 Hornet* (Kingston, Canada: Centre for International Relations, Queen's University, 2000), 13–20.

102. Horner, "Address," 145.

103. Baldassare Catalanotto and Hugo Pratt, *Once upon a Sky: 70 Years of Italian Air Force* (Rome: Lizard Edizioni, 1994), 111.

104. Quoted in Horner, "Address," 146.

105. Catalanotto and Pratt, *Once upon a Sky,* 111–12.

106. *GWAPS: Statistical Compendium and Chronology,* table 81, p. 317.

107. *GWAPS: Operations and Effects and Effectiveness,* 20.

108. *GWAPS: Logistics and Support,* 285.

109. Ibid., 217–18, 292–96.

110. *GWAPS: Weapons, Tactics, and Training,* 314–15.

111. *GWAPS: Logistics and Support,* 200.

112. Wratten, statement in "Britain and the 1991 Gulf War Witness Seminar," 69.

113. *GWAPS: Weapons, Tactics, and Training,* 311.

114. Hine, "Despatch"; *GWAPS: Logistics and Support,* 179, 200. Note that the Kuwait Air Force did not employ air refueling for its F-1s and A-4s.

115. Calculated from data in *GWAPS: Statistical Compendium and Chronology,* table 64, pp. 232–34.

116. *GWAPS: Statistical Compendium and Chronology,* table 206, pp. 653–54.

117. *GWAPS: Statistical Compendium and Chronology,* table 205, p. 651.

118. President George H. W. Bush, "Remarks at the USAF Academy Commencement Ceremony, May 29, 1991," in *Public Papers of the Presidents of the United States: George H. W. Bush,* book 1 (Washington, DC: US GPO), 576–77. For the record, Bush's second and third lessons were the value of "stealth," or low-observable, counter-radar technology, and the value of ballistic missile defense.

Operation Allied Force as a Catalyst for Change

Toward Intensified Multinational Cooperation

Maria E. Burczynska

Following the end of the bipolar power competition between East and West in 1991, European militaries, including their air forces, underwent a significant process of transformation characterized by two major trends—concentration and transnationalization.[1] These trends resulted in the creation of not only smaller but also more professional and specialized air forces, which started cooperating at an international level to a much larger extent.[2] As a result, they have become more interconnected and more interdependent than they were during the Cold War. Another trait characteristic of the post–Cold War period has been the increased number of multinational operations European air forces have been involved in as a logical result of concentration and transnationalization. Concentrated air forces became increasingly more reliant on supplementing their available capabilities through cooperation with allies. This is one of the reasons why multinational operations have become the dominant form of military intervention undertaken by European countries. Operation Allied Force (OAF), conducted in Kosovo in 1999 and the North Atlantic Treaty Organization's (NATO) first major military operation, is one example of this type of force application. It also serves as a good illustration of the challenges multinational operations present to participating air forces.

This chapter seeks to uncover the challenges that arose from the multinational nature of OAF and how they affected cooperation between the NATO countries involved as well as the operation's effectiveness. It shows that the experience of multinational operations in OAF set into motion the development of various forms of multinational cooperation within European NATO members and the alliance as a whole, further increasing the transnationalization of European air forces. Investigating the major difficulties and lessons learned from this operation, this chapter highlights in particular the issue of national caveats and the capability gap between the US Air Force (USAF) and the rest of the alliance, evident in equipment shortages and interoperability problems.

The chapter starts with a brief discussion of OAF and its significance to provide the necessary background for investigating the challenges it presented. Then it looks more closely at the concepts of military concentration and transnationalization as factors leading to enhanced multinational cooperation and discusses the major challenges encountered in OAF. Finally, it looks at how these challenges were addressed at the time of the operation and how they affected the further transnationalization of European air forces.

OPERATION ALLIED FORCE

OAF started on 24 March 1999 following the unsuccessful talks in Rambouillet, France, between Serbian leaders and the Kosovo Liberation Army.[3] The operation's objective was to stop the genocide committed by the forces of Yugoslavian president Slobodan Milosevic against the citizens of the Kosovo Province.[4] OAF's approach was characterized by the intention to gradually extend the range of targets during three phases in order to achieve its stated objectives and to coerce Milosevic to capitulate.[5] Following seventy-eight days and 38,004 sorties flown, the operation finished on 10 June 1999 after Milosevic's capitulation and agreement to withdraw Serbian forces from Kosovo.[6]

All of the then nineteen NATO countries officially supported the intervention, but they contributed to the operation to different extents.[7] For example, the Czech Republic, Greece, Hungary, Iceland, Luxembourg, and Poland did not provide any aircraft or combat forces. Their input was minimal because only Greece and Hungary contributed to the military effort by providing basing locations and the right to fly over their territories.[8] The most sizeable contributions to OAF were made by the United States, France, the United Kingdom, the Netherlands, Italy, and Germany. Their parts in the operation involved flying approximately 29,000 (United States), 2,414 (France), 1,950 (United Kingdom), 1,252 (the Netherlands), 1,081 (Italy), and 636 (Germany) sorties.[9] The three largest aircraft contributors were the United States, France, and the United Kingdom, with the Americans providing more than 700 aircraft out of the total 1,055 deployed by the alliance.[10] Furthermore, the United States, France, and the United Kingdom were also the only alliance members who contributed precision-guided munitions (PGMs).[11]

The extent to which OAF was an unambiguous air power success or was won by air power alone has been widely debated.[12] On the one hand, the operation clearly succeeded inasmuch as it achieved its objectives and coerced Milosevic to withdraw from Kosovo. On the other hand, as General Wesley Clark, US Army, pointed out later, "though NATO had succeeded in its first armed conflict, it didn't feel like a victory."[13] What he meant was that it was not entirely clear if it was solely the air operation that caused Milosevic to capitulate.[14] Whatever the rights or wrongs of this debate, OAF was certainly a significant operation for two reasons.

First, OAF justified NATO's continuing existence, which was questioned after the Soviet Union—the opponent the alliance was created to defend against in the first place—had ceased to exist. OAF was NATO's first major military operation and, as such, marked its transformation from a predominantly defensive organization into one that took up offensive activities.[15] Second, as demonstrated further in the chapter, OAF revealed significant challenges to cooperation within the alliance, which determined the future course for development of various multinational initiatives enhancing NATO's collective capability and capacity. These challenges can be grouped into two main categories: national caveats affecting, for example, the decision-making process and a significant capability gap between the United States and the rest of the alliance, evident in interoperability issues between participating air forces and the systems they used. These challenges were to be expected in a large multinational operation. However, because the Kosovo action was NATO's first major military intervention, which, moreover, took place while many of the participating air forces were still undergoing structural and organizational changes, it made clear which areas needed to be addressed to improve the effectiveness of future operations. Therefore, as this chapter shows, OAF initiated the further transnationalization of NATO air forces, especially in Europe.

POST–COLD WAR TRANSNATIONALIZATION OF AIR FORCES AND ITS CONSEQUENCES

The end of the Cold War bipolar order initiated a military transformation of NATO and some former Warsaw Pact air forces that effected their growing engagement in multinational operations. This process took the form of concentration and transnationalization, leading to a change in the quantity and quality of the affected forces. At this point, it is vital to mention that both processes had an impact predominantly on the European air forces. The USAF, although it also underwent certain changes, retained its full spectrum of capabilities.

The concentration of European air forces was a logical consequence of the end of the Cold War, which at the time seemed to designate the threat of large-scale, conventional conflict in Europe as a relic of the past. With a dramatically changed security environment, many European states stopped prioritizing large-scale conventional military capabilities, decreased their military expenditure, and started the process of reducing existing forces.[16] For example, in the United Kingdom, that process was initiated with two defense reviews published by the Conservative government: *Options for Change* (1990–1991) and *Front Line First* (1994), addressing predominantly the reduction of defense budgets. Table 10.1 illustrates the reduction in personnel in the air forces of the six major contributors to OAF.

However, military concentration did not merely mean a decrease in the size of military forces. Its ultimate goal was the creation of compact and specialized professional units that seemed more relevant to the new security environment.[17] Another characteristic of that process was the contractorization of the militaries, which involved outsourcing certain services to the private sector rather than having them delivered by the military or the civil service.[18] In the United Kingdom, this move was initiated in the 1980s and is still being continued today, involving an array of activities from office support to support for expeditionary military operations—for example, in Afghanistan.[19] Therefore, concentration should not be viewed as a decline. Although the transformed armed forces had fewer resources and personnel because of this transformation, as Anthony King argues, they emerged "qualitatively

Table 10.1.   Total Air Force Personnel of Major OAF Contributors, 1990 and 1999

Country	Personnel in 1990	Personnel in 1999
United States	571,000	361,400
France	93,100	76,400
Germany*	106,000	76,400
Italy	79,600	61,900
Netherlands	17,400	11,980
United Kingdom	89,600	55,200

* Former German Federal Republic.

Sources: Data compiled from "NATO and Non-NATO Europe," *The Military Balance* 99, no. 1 (1999): 30–103; "The Alliances and Europe," *The Military Balance* 90, no. 1 (1990): 44–96; "United States," *The Military Balance* 99, no. 1 (1999): 12–29; and "United States," *The Military Balance* 90, no. 1 (1990): 12–27.

different" from their pre-1990 counterparts as they benefitted from targeted investment in specific prioritized areas.[20] As a result, armed forces that had focused on mass and in many cases were maintained through universal conscription evolved into smaller, professional, and more specialized forces. These forces were better suited to operations at the lower end of the conflict spectrum, such as peacekeeping, peace support, and humanitarian interventions requiring deeper cooperation between military and civilian sectors than the "traditionally" perceived military functions.[21]

For the European air forces, military concentration led to the creation of specialized and more effective units, but it unfortunately did not allow for building or maintaining a full spectrum of air power capabilities. Certainly, no European air force could compete with the USAF as the most powerful air force in the world. Therefore, with the budgetary cuts, reorganization, and personnel reduction, investment in one capability was often made at the expense of another. A logical solution to this problem was increased cooperation among European air forces, which allowed individual states to make up for shortcomings in certain areas and to maintain military capability. Such multinational engagement ultimately led not only to increasing interoperability but also to interdependence, resulting in what King calls "military transnationalization."[22]

Multinational military cooperation was not, however, a new phenomenon in the post–Cold War period. States had already cooperated in the military sphere and worked in ad hoc coalitions for centuries. Nevertheless, from 1991 on, cooperation started taking place more frequently across the whole military structure, not only at the strategic level but also at the operational and tactical levels, leading to the creation of truly multinational forces within coalitions. As such, this cooperation also uncovered previously unknown challenges and difficulties, potentially disrupting the smooth running of missions and affecting the effectiveness of operations.

CHALLENGES FACED IN OPERATION ALLIED FORCE

Multinational operations are complex undertakings. They involve forces from different nations coming from different cultural backgrounds, representing different approaches, and bringing their own not always compatible equipment, procedures, and regulations to the coalition. OAF was NATO's first major military intervention conducted in a changing security environment. As such, it faced significant challenges arising from multinational cooperation as a result of national caveats and a capability gap in particular.

National Caveats

According to NATO's definition, national caveats are "any limitation, restriction or constraint imposed by a nation on its military forces or civilian elements under NATO command and control or otherwise available to NATO, that does not permit NATO commanders to deploy and employ these assets fully in line with the approved operation plan."[23] The imposition of national caveats by individual states is the norm in multinational operations for various reasons. First and foremost, they are imposed as a means of control over deployed national forces in order to minimize the costs and risks they will be exposed to.[24] Second, caveats often reflect specific domestic political considerations of nations involved in a coalition and multinational operations. As such, national caveats might represent, for example, a compromise between political parties with different views on the country's involvement in a particular operation or a way to ensure that the activities performed by the deployed forces will not damage national interests or the country's or its leaders' image, both domestically and internationally, through negative publicity.[25] Third, following from the previous point, national caveats may be imposed according to what foreign-policy behavior is considered appropriate or desirable by individual states in various conditions. Therefore, national caveats are also often linked to a nation's cultural and historical background—the values and perceptions shared by a society as a whole.[26]

As Anthony Cordesman points out, political constraints as a result of national caveats rather than capability shortcomings were "the most serious single problem" in the conduct of OAF.[27] The restraints posed by national caveats on the effectiveness of missions were visible in the process of decision making, especially decisions pertaining to the approval of targets. However, they were also reflected in the attitude that the involved countries had toward the air strikes and military intervention in general as opposed to the option to solve the conflict by diplomatic means.

In the run-up to OAF, there was a noticeable trend for alliance members geographically closest to the area of operations to be the most reluctant about a military solution to the conflict, the escalation of air strikes once the operations had started, and the option of deploying ground troops.[28] For example, the states most determined to find a nonmilitary solution to the conflict were Italy and Greece. Considering that these countries are direct neighbors of Yugoslavia, such a stance should not have come as a surprise. In the case of a military intervention, both countries would be the first to be troubled by its effects—for example, large numbers of refugees seeking shelter from the conflict.[29] The stance represented by Italy and Greece was an explicit example of their leadership's efforts to protect their national interest, in this case to maintain regional stability, while at the same time avoiding political marginalization and remaining viable members of NATO.[30] Both Italy and Greece ultimately supported the operation but also continued pushing for a peaceful solution and restraint, even once the air strikes commenced. The Italian government continued its efforts to restart the negotiations in order to resolve the conflict through diplomatic means, and the Greek leadership succeeded in convincing Milosevic to put forward a cease-fire offer for the duration of the Orthodox Easter. The coalition rejected the cease-fire offer, however, as not reliable, and the air strikes did not stop.[31] The Italian government's stance presents a particularly interesting case considering that, after all, Italy in practice was one of the major contributors of personnel and aircraft to the operation.

Germany's motivation for supporting the operation was the wish to strengthen its position in the international arena. The desire to improve its international image by standing firm against genocide and ethnic cleansing in Kosovo was influenced by Germany's past and the atrocities committed in its name during the Second World War.[32] At the same time, precisely because of its military past, Germany shared Italy and Greece's determination to resolve the conflict by diplomatic means and to minimize the use of lethal force, and so it decisively stood against escalation and the introduction of ground forces.[33]

A contrasting approach and attitude toward intervention in Kosovo was taken by the United Kingdom, which was by far more assertive, or "hawkish," as Olivier Schmitt describes it.[34] To strengthen its position within the alliance and its relationship with the United States, the United Kingdom not only lobbied firmly for a military intervention to coerce Milosevic into stopping the atrocities but also was one of the few NATO members that insisted on not ruling out the ground option. This stance was very much in line with the United Kingdom's *Strategic Defence Review* of 1998, which set the direction for transforming the British armed forces into highly deployable expeditionary units able to respond to a series of new threats and forms of conflict, including humanitarian interventions.[35] It was also concurrent with Tony Blair's Doctrine of the International Community, in which he advocated for the need of multinational cooperation and international response to the arising crises in the Balkan region.[36]

National interest also dictated the support of OAF by some smaller countries, especially the newest members of NATO—namely, the Czech Republic, Hungary, and Poland. In spite of these countries' minimal practical contribution, expressing support for the operation was extremely important for them because it was an opportunity to demonstrate that they were useful and committed members of the alliance as well as valuable allies of the United States.[37] These examples and the preceding ones make it clear that national interests and domestic politics played a major role in shaping the way individual NATO members approached the operation.

Once OAF commenced, some countries' reluctance vis-à-vis their engagement in a fully-fledged air campaign was also mirrored in the process of target selection. As General Wesley Clark pointed out later, European members of the alliance, in contrast to the United States, on the whole were less willing to strike strategically sensitive targets, such as communication lines, bridges, radio and television stations, and the power grid, because they wanted to avoid antagonizing the Serbian population and causing excessive destruction in the country.[38] Such an approach was represented especially by the French, who out of all nineteen NATO members most often used their right to veto air strikes against critical infrastructure in Serbia as well as against targets located in Montenegro.[39] Lack of unanimous decision on target approval certainly caused a great deal of friction in the alliance. It had the potential for serious consequences—for example, by endangering allied forces as a result of the inability to destroy surface-to-air missiles in Montenegro because of repetitive French vetoes.[40]

The process of target authorization was further complicated by the presence of parallel command structures in OAF. Alongside the NATO chain of command, there were national chains of command, which complicated decision making. For example, after a target was approved by General Clark at Supreme Allied Command Europe, it also had to go through a similar review in all participating states.[41] This procedure not only complicated target approval but also caused confusion regarding command and control as well as responsibility for performing a particular task.[42] Parallel decision making also had another downside. As General Clark noted later, the prolonged process of target approval often involved the public discussion of potential targets, thereby making them known to Serbian forces, which obviously hampered the effectiveness the of air strikes.[43]

Although potentially destructive for the alliance and certainly frustrating for the participating forces, friction resulting from national caveats did not ultimately destroy coalition coherence and unity during OAF. As the operation progressed, the alliance learned from the process and set out immediately to implement lessons identified. For example, by the end of the first week of the operation, the United States, the United Kingdom, France, Germany, and Italy agreed on a list of points to be used as guidance for the United States on target authorization, which, once followed, would prevent the European allies from using their veto.[44] As a compromise, targets in Montenegro were excluded unless they posed a direct threat to allied forces. Any targets within 5 nautical miles (9.3 nautical kilometers) of Belgrade required approval, as did air strikes that might incur significant civilian casualties or that targeted the power grid.[45] Although this guidance was prepared and agreed on only after OAF had started, it was able to take into account the participating states' major national caveats and therefore limited friction within the alliance regarding target authorization.

Nevertheless, the introduced guidelines did not completely prevent collateral damage, as the allies wished. As reported, OAF included more than thirty cases of unintended damage due to identification and targeting errors, twelve of which involved civilian casualties.[46] These twelve were widely publicized in the media and had an enormous effect on the conduct of the operation. The targeting errors included, for example, mistaking civilian vehicles containing refugees for a military convoy on the road between Djakovica and Decane in southeastern Kosovo, accidentally bombing a bridge in southern Serbia while a passenger train was crossing it, and bombing the Chinese embassy in Belgrade as a result of mistaking it for a building used for military purposes.[47] The bombing of the Chinese embassy probably had the most adverse effect on the conduct of the operation because it not only led to stopping any bombings in Belgrade for two weeks but also caused a diplomatic crisis between the United States and China.[48] In general, these targeting errors had a twofold effect. First, they directly influenced the introduction of very strict rules of engagement, as discussed earlier. Second, they contributed to the unrealistic expectations set for the operation. As Benjamin Lambeth points out, the goals of zero casualties and no unintended damage to nonmilitary infrastructure were adopted as almost a measure of the operation's and hence air power's success—the perceived perfectly precise tool underwent a stringent judgment under unrealistic standards.[49]

The Capability Gap between the United States and European Allies

The operation in Kosovo revealed a significant capability gap between the USAF and the air forces of European NATO members. This gap was especially visible in such areas as PGMs and command, control, and communications systems, but also in air transport and air-to-air refueling capabilities.

As a result of military concentration, European air forces had been significantly reduced in size, which had further increased the capability gap between them and US air forces. Table 10.2 illustrates the number of active personnel and aircraft of the United States and European NATO members in 1999.

The table shows that at the time of OAF the USAF had approximately 70 percent of the personnel and 81 percent of the aircraft inventory, compared to the resources possessed collectively by the European NATO members. It thus vividly illustrates the capability and capacity gap. The numbers given for the European air forces include a large variety of different types of aircraft that vastly differed in quality—several different types and generations of combat aircraft and helicopters that were not always compatible, such as Soviet Su-22s or MiG-29s on the one hand and F-16s or different versions of

Table 10.2.   NATO Air Forces Personnel and Inventory Strength, 1999

Country	Active Personnel	Total Number of Aircraft
Belgium	11,500	221
Czech Republic	15,400	291
Denmark	4,700	110
France	76,400	984
Germany	76,400	837
Greece	30,170	611
Hungary	11,500	344
Iceland	—	—
Italy	61,900	633
Luxembourg	—	—
Netherlands	11,980	304
Norway	6,700	142
Poland	55,300	650
Portugal	7,445	150
Spain	29,100	558
Turkey	63,000	775
United Kingdom	55,200	1,023
European allies total	**516,695**	**7,633**
United States	361,400	6,178

Sources: Data compiled from "NATO and Non-NATO Europe," *The Military Balance* 99, no. 1 (1999): 30–103, and "United States," *The Military Balance* 99, no. 1 (1999): 12–29.

Tornado aircraft on the other.[50] At the same time, the number of available European transport or reconnaissance aircraft was low, and only a few European air forces had any refueling capability or PGMs.[51] The type and quality of platforms and their compatibility are more important than quantity because they affect the effectiveness of an operation, especially when it is conducted in a multilateral setting. Add to these variables the differences in doctrine, training, and procedures, and it is easy to see that European NATO air forces, even with an impressive quantity of personnel and equipment, did not amount to a joint "European Air Force" with full-spectrum capabilities.

In OAF, the United States contributed the vast majority of material resources. The USAF, as already mentioned, provided more than 700 out of 1,055 aircraft and 23,315 (83 percent) of all precision and nonprecision munitions used in the operation.[52] The largest contributors of aircraft from among the European air forces were France, the United Kingdom, and Italy.[53] However, the other European allies' contributions should not be diminished because in addition to some aircraft they provided crucial basing facilities for American assets. In total, the non-US coalition allies flew approximately 40 percent of the missions.[54] Some European allies proved extremely valuable in specific mission areas, making up for other shortcomings. For example, Germany and Italy could not contribute PGMs because they did not possess such capability, but they excelled in the suppression of enemy air defenses.[55]

One of the major shortcomings revealed during OAF was the limited access to PGMs. Considering the concerns expressed by some of the coalition members about limiting offensive operations and minimizing the number of casualties, use of PGMs would be crucial to meeting those objectives. However, during the operation PGMs constituted 35 percent of all the weapons used, but only three of the participating states (the United States and to a lesser extent France and the United Kingdom) were able to deliver them, which placed a strain on the USAF.[56] Moreover, after the operation had commenced, it became evident that many of the deployed European fighter aircraft presented very little operational usefulness because they could not conduct strikes with PGMs and rarely were able to operate in all-weather or night conditions.[57] These disparities only stressed the size of the capability gap between the United States and the rest of the alliance.

Another area that revealed a significant limitation of European air forces' capabilities was air-to-air refueling (AAR). Again, most of the tanker aircraft, more than 170, were provided by the United States, whereas the European allies deployed only 13.[58]

Considering that 21 percent of all sorties performed in OAF were AAR related, the number of specialized aircraft contributed by European air forces seemed almost insignificant. Furthermore, the cooperation in the AAR area was also disrupted by interoperability issues between the UK tankers and some of the US aircraft. Undoubtedly, the Royal Air Force provided a significant share of AAR in OAF, as illustrated by the fact that 85 percent of the fuel the British tankers supplied were received by non-British aircraft.[59] The United Kingdom was also the sole NATO member that had the Joint Tactical Information Distribution System installed in their tanker aircraft and therefore proved to be one of the most interoperable AAR fleets in the operation.[60] Nevertheless, refueling turned out to be problematic between the British and American aircraft. According to Air Vice Marshal Steve Nicholl, the

A US Navy EA-6B Prowler refuels from a USAF KC-135R Stratotanker during a NATO OAF mission on 12 May 1999. US Department of Defense photograph by Senior Airman Greg L. Davis, USAF, at https://archive.defense.gov/photos /newsphoto.aspx?newsphotoid=2147.

Royal Air Force was able to refuel the US Navy and the US Marine Corps aircraft but was not compatible with the USAF platforms except for the F-16s, which are widely sold among other nations and therefore present fewer interoperability problems.[61] Considering that the United States provided the largest number of aircraft to the operation and that the United Kingdom was one of the major AAR providers, this incompatibility presented a serious interoperability issue.

Similar limitations were present in the areas of airlift, electronic-warfare assets used in support of the suppression of enemy air defenses and delivered mostly by the United States, as well as reconnaissance, especially the lack of a system for intelligence gathering and processing independent from that operated by the United States.[62] The shortcomings in the area of strategic airlift were partially mitigated by the United Kingdom, for example, through the use of commercial assets.[63] Nevertheless, following the end of the operation, European allies identified that particular capability as one of the most urgent limitations of European air power.

The capability gap in providing reconnaissance was partially filled by the deployment of US and European unmanned aerial vehicles (UAVs) to the operation, which provided the allied forces with precise, real-time reconnaissance and surveillance information. UAVs were also used in the process of target approval, cross-checking the information on potential targets gathered by the alliance's pilots.[64] Among the European contributors, both France and Germany deployed UAVs to OAF that performed 37 percent of the unmanned sorties. However, these UAVs could fly only very short distances, so the amount of reconnaissance data they could gather was significantly reduced.[65] Nevertheless, the introduction of unmanned platforms to OAF significantly improved the quality of reconnaissance missions performed. However, the lack of an intelligence-gathering and processing system independent from the US system presented a significant problem for the European allies. Because they did not possess such capability, they were completely reliant on the information supplied by the United States. After OAF concluded, several European nations, including France, Germany, and Italy, undertook steps to improve that particular capability by expanding their UAV fleets and cooperating on military communication satellite programs (France and Germany).[66]

The capability gap revealed during OAF was not only a matter of differences in available technology but also a serious matter of interoperability. When OAF was launched in 1999, members of NATO, with the exception of the three central European countries that had joined recently, already had five decades of experience in joint training and cooperation toward greater standardization and interoperability. For example, in preparation for a conflict with the Warsaw Pact, the allied forces trained together annually in Reforger, an exercise aimed at building NATO's ability to deploy its forces to West Germany.[67] Moreover, some NATO air forces also regularly took part in the Red Flag exercises organized by the USAF.[68] Therefore, one might have expected that the integration of forces in OAF would run seamlessly. It became clear, however, that experience in peacetime training and joint exercises did not always easily translate into the smooth conduct of an actual military operation. The most serious interoperability issue uncovered in OAF was related to communications capability. The problem was twofold: first, there was no common communication and information-exchange network that could be used by the United States and other NATO members, and, second, the existing communication systems used by individual allies were not compatible.[69] This incompatibility affected not only the smooth and timely exchange of information but also the security of data transmission. This particular shortcoming hampered cooperation at all levels—strategic, operational, and tactical. The existing secure communication systems were often not suited to sending large amounts of data and got easily overloaded. For example, NATO's Limited Operational Capability for Europe system, which was used in OAF for forwarding air tasking orders, could send only a very limited amount of data.[70] The lack of secure channels, such as telephone lines and radio frequencies, in addition to the limited capacity of the existing systems, forced the participating air forces to use nonencrypted channels to forward sensitive information, which increased the possibility of interception by Serbian forces. The other option was to deliver sensitive information in writing for the rest of the allies to enter manually into their national systems and databases, which would unnecessarily prolong the delivery and processing of information.[71] Both solutions significantly reduced the effectiveness of the operation, especially in relation to mobile targets, which could be in a completely different location by the time the information was processed and an air strike launched.

Cooperation in OAF was disrupted by national caveats as well as by a significant capability gap between the USAF and European NATO members, which was reflected in the lack of crucial equipment and in the noninteroperability and noncompatibility of the existing national systems. These major limitations served as a catalyst for developing efforts aimed at overcoming them once OAF was completed.

SETTING THE COURSE FOR IMPROVING THE TRANSNATIONALIZATION OF EUROPEAN AIR FORCES

OAF, as NATO's first major military intervention, set the course for addressing issues that had resulted from the transnationalization of European air forces and ensuring their compatibility. The issue that raised the biggest concerns after OAF was the aforementioned capability gap between the USAF and the rest of NATO. It was identified as the most urgent problem to deal with, and, hence, the development of the Defense Capabilities Initiative (DCI) was prioritized immediately.[72]

The DCI was approved at the Washington summit in April 1999 and aimed at bridging the capability gap between NATO members and improving their interoperability.[73] These objectives were to be achieved through focusing on areas such as deployability, mobility, sustainability, and survivability of allied forces as well as on areas such as logistics, effective management, and command and control.[74] The DCI identified fifty-eight goals toward upgrading military capabilities. Some goals, such as training in increased cooperation and coordination between allies, were considered relatively easy to implement and cost effective. Others, such as modernization of existing equipment, required greater investment of time and money.[75] The High Level Steering Group was created at the summit and put in charge of overseeing the implementation of the DCI. One of the substantial improvements leading toward achieving the DCI's objectives and recognized by the steering group was the introduction of the concept of a Multinational Joint Logistics Center, aimed at the creation of an integrated allied system for theater logistics.[76]

OAF made it unambiguously clear that some of the areas identified in the DCI needed to be prioritized in order to work toward closing the capability gap between the United States and Europe. All of the set objectives initially had equal priority. However, after OAF had concluded, it became clear that two of them were more urgent for the European allies—first, forces' deployability and sustainability and, second, the allies' agreement on common goals in defense expenditure.[77]

The DCI presented promising opportunities for the improvement of cooperation within NATO as an alliance. All of the allied states understood that the DCI was a long-term commitment. However, in spite of this common understanding, they had quite different perceptions of how it should be conducted. For example, because of its focus on the concept of the "Revolution in Military Affairs" and the promotion of advanced equipment, the initiative was criticized as being affordable only for the more powerful European members of NATO, such as the United Kingdom, France, and Germany, and thus leaving out the smaller, less-wealthy states.[78] Those smaller states, especially the new central and eastern European members, perceived the DCI as a long-term goal that should be implemented after their militaries had adapted to match the standards set by NATO.[79] The perceptions of the DCI harbored among the larger European NATO countries also varied. For example, Germany had a very selective approach to the set goals and was interested in working toward only a few of them—namely, strategic airlift, command and control, and intelligence. In contrast, the United Kingdom and France considered the initiative's objectives as confirmation of the course of military restructuration and modernization they had already adopted.[80] The United States, in turn, perceived it as a solution to upgrade the European armed forces and to close the capability gap.[81] Those differing perceptions and the imposition of national caveats interrupted multinational cooperation and, as a result, affected the initiative's success. Certainly, the DCI correctly identified areas that needed work to bridge the gap between the United States military and European militaries. Nevertheless, it did not contribute significantly to the development of the necessary capabilities.[82] National caveats about the utility of military force and the kind of armed forces required by the individual European NATO members were therefore an obstacle in the successful implementation of the DCI.

When OAF had been completed, the allied air forces commenced some efforts to address the revealed technological gap. Those countries that could afford it started to build their own high-tech systems. For example, to address the PGM shortage, France began developing the Advanced Air-to-Surface Missile.[83] However, the majority of allies, aware of budgetary constraints, opted to address the capability gap by intensifying their participation in various cooperative initiatives and programs in order to pursue the goals set by the DCI. For example, after OAF, Belgium, Denmark, Italy, the Netherlands, Norway, Turkey, and the United Kingdom agreed on the procurement of the Joint Direct Attack Munition to make up for the PGMs shortcoming.[84] The Joint Direct Attack Munition is a kit that facilitates the conversion of nonguided munitions into PGMs, and because it uses the Global Positioning System, it also upgrades them with all-weather capability.[85] It therefore offered a much less expensive option of acquiring PGM capability by using already-existing resources in place of developing entirely new technology from scratch.

A priority area in improving the cohesion and integration of NATO air forces was to ensure the unification of policies and procedures, especially in the area of C4—command, control, communications, and computers.[86] Viewed as equally important for future operations were unified tactics, agreement on the military or diplomatic means used to resolve the particular conflict, and the perception of strategic objectives—all of which should be clearly set out in allied publications. Hence, the importance of continuing to write allied joint publications and establishing a common NATO doctrine was also stressed.[87] Setting overarching standards for all NATO members was perceived rightly as the way to ensure interoperability and to minimize friction among participants in any future operation.

Finally, as pointed out in the US Department of Defense's report to Congress in 2000, "Operation Allied Force also validated the need for joint, integrated training among the Services to enhance their ability to execute both, joint and coalition air operations."[88] Although this statement referred to the US armed forces specifically, it may be perceived as a general lesson from OAF applicable to all NATO member states and their air forces. Coalition building requires interoperability among its members, which comes, first, from standardized procedures and, second, from experience working together.

The cooperative multinational initiatives pursued under DCI were important for European air forces, especially the smaller ones. Many European states could not afford to buy significant amounts of new equipment or, what would have been even more expensive, to develop their own systems to make up for existing limitations and to build a fully capable air force. The majority therefore resorted to various multinational initiatives based on the idea of pooling and sharing resources, which offered a (cost-)effective way of accessing necessary equipment and boosting operational capability.[89]

Conclusion

OAF can undoubtedly be perceived as a milestone in the history of NATO's multinational operations. First, it validated the very existence of the alliance and set a course for its future engagements. Not only was Allied Force the first major military operation conducted by NATO, but it was also carried out in the new, post–Cold War security environment, where the alliance's prime opponent and the very reason for its establishment had ceased to exist. The operation and its outcome did not live up to the overly optimistic initial hopes of some that it could be concluded within days but did prove that the alliance was still an important organization with much potential to be used in resolving conflicts and in answering crises arising in the challenging post-1990 reality.

Second, OAF highlighted the serious challenges that multinational cooperation may present. A lack of compatible equipment, standardized procedures, and parallel command structures as well as an overreliance on US resources disrupted the conduct of the operation and highlighted the urgency of improving interoperability within the alliance. The operation in Kosovo not only identified areas of serious limitations in capability and capacity, especially among the European allies, but also made it clear that NATO in its current shape, not to mention its individual members, would not be able to oppose a peer competitor. The situation in Kosovo was characterized by asymmetry between the sides in the conflict, wherein the alliance, despite all of the interoperability issues, had an undeniable advantage over Milosevic's forces. In a conflict with a more equal opponent, the capability gap, the interoperability issues, and the overly restraining national caveats would have had a significantly more adverse effect on the operation and its outcome. Therefore, it can be asserted that the drive toward multinational cooperation and the need to address these challenges have regained particular importance recently, when the threat of state-on-state conflict has again been elevated to the political agenda as a result of the rise of China and Russia.[90]

The process of military transformation taking place among the armed forces and shrinking defense budgets ensured that multinationality became the dominant form of post–Cold War military interventions and that the issues identified during OAF needed to be addressed for cooperation to be effective. Increased multinational cooperation in the form of the DCI was prioritized as a solution. The DCI was developed into another related effort, Smart Defence, but the objectives remained unchanged—to build a collective military capability and at the same time to provide NATO members with a cost-effective way to access required capabilities and to gain the necessary experience in using them. However, in spite of the efforts since OAF, the capability gap between the USAF and European air forces had still not been resolved more than a decade later, as demonstrated by the problems experienced during Operation Unified Protector in Libya in 2011.[91]

NOTES

1. The chapter was written with financial support from the Royal Air Force Centenary PhD Bursary in Air Power Studies awarded to the author by the Royal Air Force Museum. The author expresses her deepest gratitude to Dr. Bettina Renz for her continuous support and irreplaceable advice given while the chapter was being written.

2. Anthony King, *The Transformation of Europe's Armed Forces: From the Rhine to Afghanistan* (Cambridge: Cambridge University Press, 2011), 32–44.

3. John E. Peters, Stuart Johnson, Nora Bensahel, Timothy Liston, and Tracy Wilson, *European Contributions to Operation Allied Force* (Santa Monica, CA: RAND, 2001), 9–17.

4. Benjamin S. Lambeth, *NATO's Air War over Kosovo: A Strategic and Operational Assessment* (Santa Monica, CA: RAND, 2001), 19.

5. See Dick Leurdijk and Dick Zandee, *Kosovo: From Crisis to Crisis* (Aldershot, UK: Ashgate, 2001), 74–77, and Lambeth, *NATO's Air War over Kosovo*, 19–43.

6. Anthony H. Cordesman, *The Lessons and Non-lessons of the Air and Missile Campaign in Kosovo* (Westport, CT: Praeger, 2001), 53.

7. The NATO countries were Belgium, Canada, Czech Republic, Denmark, France, Germany, Greece, Hungary, Iceland, Italy, Luxembourg, the Netherlands, Norway, Poland, Spain, Turkey, the United Kingdom, and the United States (Bruce R. Nardulli, Walter L. Perry, Bruce R. Pirnie, John Gordon IV, and John G. McGinn, *Disjointed War: Military Operations in Kosovo, 1999* [Santa Monica, CA: RAND, 2002], 26).

8. Derek S. Reveron, "Coalition Warfare: The Commander's Role," in *Immaculate Warfare: Participants Reflect on the Air Campaigns over Kosovo and Afghanistan,* ed. Stephen D. Wrage (Westport, CT: Praeger, 2003), 59.

9. Peters et al., *European Contributions,* 18–23.

10. Ibid., 23.

11. Olivier Schmitt, *Allies That Count: Junior Partners in Coalition Warfare* (Washington, DC: Georgetown University Press, 2018), 86, 92; Peters et al., *European Contributions,* 23–24.

12. See, for example, Daniel A. Byman and Matthew C. Waxman, "Kosovo and the Great Air Power Debate," *International Security* 24, no. 4 (2000): 5–38; Ivo Daalder and Michael O'Hanlon, *Winning Ugly: NATO's War to Save Kosovo* (Washington, DC: Brookings Institution Press, 2001); Lambeth, *NATO's Air War over Kosovo;* Lawrence Freedman, "Victims and Victors: Reflections on Kosovo War," *Review of International Studies* 26, no. 3 (2000): 335–58; Leurdijk and Zandee, *Kosovo;* Martin Aguera, "Air Power Paradox: NATO's Misuse of Military Force in Kosovo and Its Consequences," *Small Wars and Insurgencies* 12, no. 3 (2001): 115–28; William M. Arkin, "Operation Allied Force: 'The Most Precise Application of Air Power in History,'" in *War over Kosovo: Politics and Strategy in a Global Age,* ed. Andrew J. Bacevich and Eliot A. Cohen (New York: Columbia University Press, 2001), 1–37.

13. Wesley K. Clark, *Waging Modern War: Bosnia, Kosovo, and the Future of Combat* (Oxford: PublicAffairs, 2001), 421.

14. See Lambeth, *NATO's Air War over Kosovo,* 67–86; Leurdijk and Zandee, *Kosovo,* 77–83; and Arkin, "Operation Allied Force," 26–29.

15. Before OAF, NATO forces were involved in several minor operations in Bosnia, such as Maritime Monitor, Sky Monitor, Maritime Guard, Deny Flight, Sharp Guard, and Deliberate Force. However, OAF was undoubtedly the first large-scale military operation conducted by the alliance. See Peter J. Anderson, "Air Strike: NATO Astride Kosovo," in *The Kosovo Crisis: The Last American War in Europe?,* ed. Anthony Weymouth and Stanley Henig (Edinburgh: Pearson Education, 2001), 185–90; Patricia A. Weitsman, *Waging War: Alliances, Coalitions, and Institutions of Interstate Violence* (Stanford, CA: Stanford University Press, 2014), 74–75.

16. Gerhard Kümmel, "A Soldier Is a Soldier Is a Soldier?! The Military and Its Soldiers in an Era of Globalisation," in *Handbook of the Sociology of the Military*, ed. Giuseppe Caforio (New York: Springer, 2006), 426.

17. King, *The Transformation of Europe's Armed Forces*, 32–40; Philippe Manigart, "Restructuring of the Armed Forces," in *Handbook of the Sociology of the Military*, ed. Caforio, 331–35.

18. Jay Edwards, *Contractorization of UK Defense: Developing a Defense-Wide Contractorization Strategy and Improving Implementation*, occasional paper (London: Royal United Services Institute, June 2018), 1.

19. Ibid., 2.

20. King, *The Transformation of Europe's Armed Forces*, 33.

21. Kümmel, "A Soldier Is a Soldier Is a Soldier?!," 427.

22. King, *The Transformation of Europe's Armed Forces*, 40–44.

23. NATO Standardization Agency, *NATO Glossary of Terms and Definitions AAP-06 (English and French)* (Brussels: NATO, 2017), 21.

24. Per M. Frost-Nielsen, "Conditional Commitments: Why States Use Caveats to Reserve Their Efforts in Military Coalition Operations," *Contemporary Security Policy* 38, no. 3 (2017): 374–79; Regeena Kingsley, "#14 an Alarming New Norm: National Caveat Constraints in Multinational Operations," *Military Caveats*, July 3, 2017, at http://militarycaveats.com/14-an-alarming-new-norm-national-caveat-constraints-in-multinational-operations; Stephen M. Saideman and David P. Auerswald, "Comparing Caveats: Understanding the Sources of National Restrictions upon NATO's Mission in Afghanistan," *International Studies Quarterly* 56 (2012): 79–80.

25. Frost-Nielsen, "Conditional Commitments," 374–79.

26. Saideman and Auerswald, "Comparing Caveats," 80–81.

27. Cordesman, *Lessons and Non-lessons*, 66.

28. Umberto Morelli, "Italy: The Reluctant Ally," in *The Kosovo Crisis* ed. Weymouth and Henig, 60.

29. Ibid., 60–61.

30. Ibid., 60–66; Reveron, "Coalition Warfare," 54.

31. Leurdijk and Zandee, *Kosovo*, 87; Reveron, "Coalition Warfare," 54.

32. Reveron, "Coalition Warfare," 55; Schmitt, *Allies That Count*, 96–97.

33. Schmitt, *Allies That Count*, 97–98.

34. Ibid., 83.

35. *The Strategic Defence Review*, CM 3999 (London: Her Majesty's Stationery Office, July 1998), paragraphs 29–31, 77–87.

36. Blair's full speech at the Chicago Economic Club outlining the doctrine can be found on the UK National Archives website. See "Doctrine of the International Community," April 24, 1999, The National Archives of the United Kingdom, at https://webarchive.nationalarchives.gov.uk/+/http://www.number10.gov.uk/Page1297.

37. Weitsman, *Waging War*, 88. See more in Paul E. Gallis, *Kosovo: Lessons Learned from Operation Allied Force*, report for Congress (Washington, DC: Congressional Research Service, 19 November 1999), 17–19.

38. Clark, *Waging Modern War*, 238–39.

39. Cordesman, *Lessons and Non-lessons*, 68–69; Peters et al., *European Contributions*, 28–29.

40. Peters et al., *European Contributions*, 29.

41. Ibid., 25–29.

42. Lambeth, *NATO's Air War over Kosovo*, 209–11.

43. Clark, *Waging Modern War*, 428.

44. Dag Henriksen, *NATO's Gamble: Combining Diplomacy and Airpower in the Kosovo Crisis, 1998–1999* (Annapolis, MD: Naval Institute Press, 2007), 21–22.

45. US Department of Defense, *Kosovo/Operation Allied Force After-Action Report*, report to Congress (Washington, DC: US Department of Defense, 31 January 2000), 24; Henriksen, *NATO's Gamble*, 22.

46. Lambeth, *NATO's Air War over Kosovo*, 136.

47. For more examples, see Daalder and O'Hanlon, *Winning Ugly*, 240–42.

48. Lambeth, *NATO's Air War over Kosovo*, 144.

49. Ibid., 139–40.

50. See "NATO and Non-NATO Europe," *The Military Balance* 99, no. 1 (1999): 30–103.

51. See ibid.

52. Lambeth, *NATO's Air War over Kosovo*, 64–66; Weitsman, *Waging War*, 84.

53. Weitsman, *Waging War*, 84.

54. Clark, *Waging Modern War*, 430.

55. Peters et al., *European Contributions*, 32.

56. Arkin, "Operation Allied Force," 21.

57. Peters et al., *European Contributions*, 35.

58. Ibid., 33.

59. UK House of Commons Defense Committee, *Fourteenth Report*, sess. 1999–2000, paragraph 165, at https://publications.parliament.uk/pa/cm199900/cmselect/cmdfence /347/34715.htm#a45.

60. Ibid.

61. UK House of Commons Defense Committee, *Minutes of Evidence*, vol. 2 of *Fourteenth Report* (HC 347-II), April 12, 2000, question 446, at https://publications.parliament .uk/pa/cm199900/cmselect/cmdfence/347/0041202.htm.

62. Peters et al., *European Contributions*, 64–65, 67; Tony Mason, "Kosovo: The Air Campaign," in *Britain, NATO, and the Lessons of the Balkan Conflicts 1991–1999*, ed. Stephen Badsey and Paul Lastawski (London: Frank Cass, 2004), 58.

63. UK Ministry of Defense, *Kosovo: Lessons from Crisis*, CM 4724 (London: UK Ministry of Defense, June 5, 2000), paragraphs 6.46, 4.48.

64. US Department of Defense, *Kosovo*, 57; Lambeth, *NATO's Air War over Kosovo*, 95–96.

65. Peters et al., *European Contributions*, 31.

66. Ibid., 67–68.

67. James P. Thomas, *The Military Challenges of Transatlantic Coalitions*, Adelphi Papers 40.333 (London: International Institute for Strategic Studies, 2000), 35.

68. Ibid., 36.

69. US Department of Defense, *Kosovo*, 49.

70. Peters et al., *European Contributions*, 56–57.

71. Ibid., 57.

72. US Department of Defense, *Kosovo*, 25.

73. NATO, "Defense Capabilities Initiative," North Atlantic Council press release, April 25, 1999, at https://www.nato.int/cps/en/natohq/official_texts_27443.htm.

74. NATO, *Defense Capabilities Initiative (DCI)*, NATO Fact Sheet, December 2, 1999, at https://www.nato.int/docu/comm/1999/9912-hq/fs-dci99.htm.

75. Hans-Christian Hagman, *European Crisis Management and Defence: The Search for Capabilities* (Abingdon, UK: Routledge, 2002), 16.

76. See more in NATO, *NATO Logistics Handbook*, 3rd ed. (Brussels: NATO, October 1997), chap. 13, "Multinational Logistics," at https://www.nato.int/docu/logi-en/1997/lo-1311.htm.

77. Peters et al., *European Contributions*, 81–83.

78. See Hagman, *European Crisis Management and Defence*, 16–18; Simen Andreas Jensen, "NATO and the Smart Defense Initiative: An Analysis in the Context of Post–Cold War Capability Initiatives in NATO," master's thesis, University of Oslo, 2014, 42–44.

79. Hagman, *European Crisis Management and Defence*, 16–17.

80. Ibid., 17.

81. Ibid.

82. Ibid., 18.

83. Peters et al., *European Contributions*, 62.

84. Ibid., 63.

85. US Navy, *Joint Direct Attack Munition (JDAM)*, US Navy Fact File, February 23, 2017, at https://www.navy.mil/navydata/fact_display.asp?cid=2100&tid=400&ct=2; Peters et al., *European Contributions*, 63.

86. US Department of Defense, *Kosovo*, 26–27.

87. Peters et al., *European Contributions*, 99.

88. US Department of Defense, *Kosovo*, 117.

89. See, for example, Maria Burczynska, "Multinational Cooperation: Building Capabilities in Small Air Forces," *European Security* 28, no. 1 (2019): 85–104.

90. HM Government, *National Security Capability Review* (March 2018), 5–6, at https://assets.publishing.service.gov.uk/government/uploads/system/uploads/attachment_data/file/705347/6.4391_CO_National-Security-Review_web.pdf.

91. See, for example, Elisabeth Quintana, Henrik Heidenkamp, and Michael Codner, *Europe's Air Transport and Air-to-Air Refuelling Capability: Examining the Collaborative Imperative*, occasional paper (London: Royal United Services Institute, August 2014), 6, at https://rusi.org/publication/occasional-papers/europes-air-transport-and-air-air-refuelling-capability-examining.

Operation Iraqi Freedom

The Allied Contribution

Benjamin S. Lambeth

As in the first Persian Gulf War in 1991, Washington's main partner in Operation Iraqi Freedom (OIF) was the United Kingdom.[1] By every measure that matters, especially given its small size and limited resources, the United Kingdom was a coequal player with the United States when it came to the quality of its equipment, military leadership, concepts of operations, and combat prowess. Such close involvement was hardly surprising; the United States and Great Britain have had a long-standing special relationship dating back to the early twentieth century. The United Kingdom had been a close partner of the United States in a succession of subsequent United Nations (UN)–approved military contingency responses, including Operation Deliberate Force and Operation Allied Force over the Balkans during the 1990s and the decade-long enforcement of the UN-mandated no-fly zones over post–Operation Desert Storm Iraq through Operations Northern Watch and Southern Watch (the British portions of which were code-named Operation Resinate), the UN effort to enforce the no-fly zones over Iraq.

In addition, the Royal Air Force (RAF) had routinely trained with the US Air Force (USAF) in realistic large-force exercises such as Red Flag at Nellis Air Force Base (AFB), Nevada, and similar training evolutions elsewhere around the world. Perhaps most important of all, ever since the terrorist attacks of 11 September 2001, the British Ministry of Defence (MoD) had maintained an embedded senior leadership and staff presence at US Central Command (CENTCOM) headquarters at MacDill AFB, Florida, in connection with Operation Enduring Freedom (OEF), to which the RAF contributed notably in the support role by providing VC-10 and Tristar tankers, E-3D Sentry Airborne Warning and Control System (AWACS) and Canberra PR9 reconnaissance aircraft, and intratheater airlifters as well as by basing provisions at the British Indian Ocean island Diego Garcia that were crucial for supporting US bomber operations.[2]

Yet the United Kingdom's participation in the second Gulf War was by no means a foregone conclusion. On the contrary, Britain's close involvement in the planning for the campaign from its earliest months continued, almost up to the beginning of combat operations, against a backdrop of persistent uncertainty as to whether British forces would actually take part in those operations.[3] Barely a day before the first bombs fell, Prime Minister Tony Blair was subjected to what was, in effect, a vote of confidence in Parliament. The vote passed by a comfortable margin of 396 to 217, largely on the strength of Blair's compelling performance in laying out the case for war.[4] After that, the British government secured parliamentary approval to use "all means necessary" in the conduct of the impending campaign.[5]

In the end, all of the required pieces fell into place in time for the United Kingdom's combat involvement, code-named Operation Telic, to commence at the war's opening moments. There was, moreover, a closer alignment of American and British campaign objectives for OIF than had been the case for enforcing the no-fly zones over northern and southern Iraq. It allowed the campaign to be conducted at the optimum tempo and with minimum political friction. After the campaign ended, the initial after-action report by the MoD affirmed that the nation's "overriding political objective [had been] to disarm Saddam [Hussein] of his weapons of mass destruction."[6] A subsequent report by Britain's comptroller and auditor general further affirmed that a second key task had been the elimination of Hussein's regime.[7] Toward both ends, the United Kingdom contributed, among other assets, some 46,000 military personnel, 19 warships, 115 fixed-wing aircraft, and nearly 100 helicopters.[8]

Australia likewise offered a spirited and substantial military contribution to OIF that reflected a deep and politically courageous national commitment. Under the successive code names Operation Bastille and Operation Falconer, the Australian government provided twenty-two aircraft—nineteen from the Royal Australian Air Force (RAAF) and three from the Australian army—and upward of 2,000 military personnel to the coalition effort. In addition to the RAAF contingent, the Australians committed a range of Royal

Australian Navy and Army assets. That contribution included a national headquarters similar to but smaller than that of the British contingent, which was colocated with CENTCOM's forward headquarters at Camp As Saliyah in Qatar.[9]

Australia's contribution to OIF stemmed from a long history of close bilateral ties with the United States, in this case going back to the Australian government's decision in 1941 to align itself formally with Washington in the security arena, an agreement that was subsequently ratified in the Australia–New Zealand–US (ANZUS) Treaty of 1951 and that has been sustained ever since by an extensive and continuing series of bilateral service-to-service relationships in such areas as joint training (the RAAF had also participated for years in the USAF's recurrent Red Flag exercises), contingency planning, intelligence sharing, and, in the cases of Korea and Vietnam, actual combat as partners in arms.

Australia's input consisted first of Operation Bastille, the forward deployment of Australian forces to CENTCOM's area of responsibility and initial area orientation and in-theater training, followed thereafter by Operation Falconer, the actual participation of Australian forces in combined coalition combat to help disarm Hussein's regime. This contribution came at a time when the Australian Defence Forces (ADF) were heavily engaged in other forward-deployed military commitments, including Operation Slipper, the ADF's involvement in the US-led OEF against the Taliban and Al Qaeda in Afghanistan, which commenced immediately after the terrorist attacks of 11 September 2001.[10]

Although the RAAF's initial involvement in early US Central Command Air Forces (CENTAF) planning workups began in late summer 2002, it was not until 10 January 2003 that the Australian government formally announced that it would deploy ADF units to the Middle East in case such a commitment should become necessary to help implement UN Security Council resolutions calling for a disarmament of Iraq—by means of force should matters come to that.[11] Three weeks later, on 1 February, the government declared that the ADF would commit a squadron of fourteen F/A-18 Hornet strike fighters as well as three C-130s (two C-130Hs and a C-130J), two AP-3C maritime patrol and surveillance aircraft, and three Australian army CH-47D Chinook helicopters, along with a forward air command element that, according to an ADF spokesman, would reside in CENTAF's Combined Air Operations Center (CAOC) and be "responsible for coordinating air operations with coalition partners and providing national control of Royal Australian Air Force assets."[12]

On 18 March, the day President George W. Bush delivered his ultimatum to Saddam Hussein and his two sons to leave Iraq within forty-eight hours or face the full force of coalition operations to drive them out of power, Prime Minister John Howard advised the Australian Parliament that his government had authorized the ADF to join in force employment by the coalition once the campaign was under way. The ADF commenced Operation Falconer immediately afterward, with Australian air, naval, and special forces initiating combat operations with the CENTCOM force components with which they had planned and trained over the preceding months.[13]

COMMAND ARRANGEMENTS

The British military contingent that since OEF had been embedded at CENTCOM headquarters began getting indications as early as May 2002 that the command was increasingly engaged in "no foreign nationals" planning, with a likely Iraq contingency in mind, because normally, in the words of Lieutenant General John Reith of the British Army, the eventual Chief of Joint Operations (CJO) for OIF, they had "very, very good access on everything."[14] The RAF in particular had embedded staff officers at all echelons both at CENTCOM's headquarters in Florida and at CENTAF's headquarters at Shaw AFB, South Carolina, as well as in CENTAF's forward-deployed CAOC at Prince Sultan Air Base, Saudi Arabia. That early embedding of key personnel at all levels ensured visibility, credibility, and the development of a deep trust relationship between the British team and its American counterparts.

Once it became clear that British involvement in such a contingency might eventually take place, the MoD established a contingency-operations group comprising representatives from the MoD's Defence Crisis Management Organization and its Permanent Joint Headquarters (PJHQ) at Northwood. It further established a current commitments team to manage all preparations that would require ministerial direction. For its part, PJHQ formed a contingency-planning team to plan the eventual British military contribution should the British government approve it. The Chief of the Defence Staff (CDS) then issued a planning directive to PJHQ that the contingency-planning team used to formulate the plan. As for the RAF's prospective involvement, headquarters Strike Command convened a contingency-action group to plan the air portion of the chosen course of action, employing cell members who were overseen by the contingency-plans division at Strike Command.

Roughly concurrently, the CDS in the autumn of 2002 appointed the Northwood-based commander of PJHQ, Lieutenant General Reith, as joint commander for all British military involvement in the impending campaign. As the designated CJO, Reith exercised operational command over all participating British forces through PJHQ and was responsible to the CDS for the conduct of operations. He delegated operational control of British forces committed to Operation Telic to the three-star national contingent commander, Air Marshal Brian Burridge, who was to be forward-deployed in the theater and to report daily to the CJO.[15]

Air Marshal Burridge was the senior MoD representative to CENT-COM's commander, General Tommy Franks, US Army, with whom he would have close daily contact throughout the campaign. He later acknowledged that Franks's forward deployment to CENTCOM's area of responsibility constituted a major improvement in command efficiency over the OEF experience.[16] Burridge reported through the CJO to the defense staff in the MoD, with Reith serving as a welcome buffer between Burridge and London. This arrangement closely followed the pattern first established by the British in Operation Desert Storm.

Reith further delegated tactical command of all RAF forces through Air Marshal Burridge to the two-star British air contingent commander, then Air Vice Marshal Glenn Torpy. Torpy's initial responsibility was to establish the force in theater; then, during the execution phase, he was to ensure that British forces were used as effectively and efficiently as possible and that operations were conducted as safely as possible in light of combat conditions within the constraints of British policy; and finally, after the period of major combat was over, he was to bring the forces home. Air Vice Marshal Torpy delegated operational and tactical control of all British air assets to the air component commander, Lieutenant General T. Michael Moseley, USAF. This delegation was consistent with years of British-American military interaction within the North Atlantic Treaty Organization (NATO) and in connection with Operations Northern Watch and Southern Watch.

Torpy's air contingent headquarters staff at Prince Sultan Air Base numbered about 200 personnel, some of whom were embedded in various CAOC staff positions. Air Commodore Chris Nickols headed the contingent's staff and also was one of three rotating one-star CAOC directors each day who oversaw the day-to-day execution of the air tasking order for General Moseley. That arrangement gave the RAF both visibility and influence within the CAOC organization. It also preserved British direction of British forces and ensured that those forces would undertake only specific operations approved by British commanders. In this post–Desert Storm arrangement, the United Kingdom built on the structure that it had developed a decade earlier in connection with its involvement with CENTCOM in enforcing the no-fly zones over Iraq.

Burridge noted that his role as the British national contingent commander focused specifically on three areas: first, supporting the three British military contingents (air, land, and maritime); second, informing the senior government leadership in London of details they needed to know for conducting responsible political and military decision making; and third, influencing CENTCOM's planning and execution of the campaign to the extent possible and appropriate. The CJO assigned different British forces to different CENTCOM missions. Burridge, as the designated wielder of operational control, was assigned forces and tasks by the CJO and, in his words, "just had to match them up with the American plan."[17] Burridge attributed the resultant smoothness of fit and flow among the various players to the key personalities involved, a fact on which the American system heavily depended.

Burridge pointed out that many on General Franks's CENTCOM staff "would regard us [the British staff] as their conscience because we see things through different eyes."[18] He attributed this good relationship to the many years of close RAF involvement with CENTCOM going back to Desert Storm and the subsequent enforcement of the no-fly zones as well as to the fact that Franks and his principal deputies recognized the quality of the thinking that the embedded British staff brought to CENTCOM's policy and strategy deliberations. That broad-based acceptance enabled the British contingent, as appropriate, to influence CENTCOM's decision making from the bottom up.

With respect to Australia's involvement in CENTCOM's command-and-control arrangements, Prime Minister Howard announced on 18 March that the Australian government had authorized Chief of the Defence Force General Peter Cosgrove to offer already deployed ADF forces as a contribution to any US-led coalition that might commit to combat operations in accordance with existing UN resolutions authorizing the use of force against Iraq.[19] Throughout Operations Bastille and Falconer, General Cosgrove retained full command of all Australian forces. Operational and tactical control, as in the case of British forces committed to the campaign, was seconded to CENTCOM's component commanders as appropriate, but national command of those forces remained with the commander of the Australian national headquarters for Middle East operations, Brigadier Maurie McNarn. The Australian Department of Defence's after-action report explained that "this arrangement let coalition commanders assign specific tasks to ADF forces while they remained under their Australian commanding officers at the unit level." In addition, "although ADF force elements worked toward the overall coalition combat plan, there were processes in place to ensure that Australian forces were always employed in accordance with Australian government policies."[20]

EARLY-PLANNING INVOLVEMENT

The British were the first allies to be brought into CENTCOM's planning for OIF. They were invited by the US government to join in the process in June–July 2002, well in advance of Australia and at a time when US forces were still reconstituting after having just completed the major combat portion of OEF in Afghanistan. After Britain's secretary of state for defence announced in September 2002 that the United Kingdom was involved in contingency planning for a possible Iraq scenario, key RAF station and squadron commanders were drawn into the planning process. Even at that early stage of preparations, a significant portion of the RAF's high-readiness forces were already deployed and flying operational missions in CENTCOM's area of responsibility as part of Northern Watch and Southern Watch. Regarding this early involvement, Burridge later recalled that although no timetable had been announced or even determined at that point, "it was put to me that if the [United Kingdom] was at any stage likely to participate, then best we at least understand the planning and influence the planning for the better."[21]

The British contingent was a fully invested participant in CENTCOM's Internal Look planning exercise in December 2002, during which CENTAF's force requirements and the RAF's contribution to the impending campaign, were it to participate, were finally nailed down.[22] The RAF also participated in multiple CENTAF "chair-fly" exercises as well as in actual large-force strike and other mission rehearsals at Nellis AFB and elsewhere. The RAF's E-3D Sentry AWACS community conducted tactical employment seminars, did spin-up training that included getting needed US combat information release and access, and conducted briefings and simulations prior to final mission certification.[23]

Shortly before the year-end holiday season in 2002, the British contingent conducted an exercise that determined that the command structure in place was, in Burridge's words, "pretty much 95 percent right."[24] In the course of that exercise, the contingent identified potential friction points, and liaison officers were installed at those points. That timely evolution bore out the value of mission rehearsals. The British government did not, however, make its final decisions regarding the composition and deployment of British assets until early 2003.

In August 2002, the US government invited initial Australian participation in CENTCOM's planning workups in Tampa, Florida, albeit with no firm military commitment sought by either side. The well-developed personal ties and trust relationships established within CENTCOM throughout the course of Australia's earlier contributions via Operation Slipper made it clear to all that the initial groundwork was now in place should Australia be included in a US-led combat coalition against Iraq.[25]

By August 2002, the ADF's joint planning staff had gained a fairly thorough understanding of CENTCOM's evolving contingency plans for Iraq and began developing appropriate options for the Australian government to consider should the latter ultimately decide that Australia would participate in coalition operations against Iraq. Later, in November, Prime Minister Howard declared that the ADF had "made appropriate contingency arrangements," including moving the Australian national headquarters for Middle East operations forward to be colocated with CENTCOM's deployed headquarters in Qatar.[26] The following month the government directed the ADF to begin training for combat operations in case Iraq failed to comply with the weapons-inspection regime stipulated in UN Resolution 1441. Finally, on 10 January 2003, the Australian government, as the start of its initial deployment of forces to CENTCOM's area of responsibility neared under the aegis of Operation Bastille, formally declared that it would commit ADF forces to the allied coalition and would begin preparing them for possible combat operations soon to come. The ADF's deployment itself finally began on 23 January 2003 with the departure of the amphibious transport ship HMAS

Kanimbla from Sydney to the North Arabian Gulf and other Australian force elements concurrently moving forward by air.[27]

THE ALLIED FORCE COMPONENT

In a force buildup code-named Operation Warrior, the United Kingdom committed to OIF the largest composite military force that it had deployed since its contribution to Operation Desert Storm. RAF Strike Command was the designated force provider for all British air assets presented for the impending campaign. In response to requests submitted by CENTCOM and CENTAF and working closely with PJHQ, Strike Command identified RAF force requirements and volunteered options, including the new Storm Shadow standoff hard-target munition and Tornado F3 interceptors, to help with the anticipated defensive counterair effort.[28]

Even before the start of focused planning for the impending campaign, the RAF already had in place a well-functioning presence of some 25 aircraft and associated personnel in the Gulf region. On 6 February 2003, Britain's secretary of state for defence disclosed that the RAF's contribution to Iraqi contingency response needs would be increased to 100 fixed-wing aircraft manned and supported by an additional 7,000 uniformed British personnel. This increased force contingent included E-3D Sentry AWACS and Nimrod and Canberra PR9 reconnaissance aircraft, VC-10 and Tristar tankers, Tornado F3 counterair fighters, Tornado GR4 and Harrier GR7 strike aircraft (with Storm Shadow missiles fitted to the GR4), and C-130 Hercules intratheater transports. The deployment reinforcement provided 115 fixed-wing RAF aircraft and about 100 helicopters, including 27 RAF Pumas and Chinooks that were fielded by Joint Helicopter Command. Among the numerous RAF ground-support assets were Rapier surface-to-air missiles, force-protection units, and explosive-ordnance disposal and expeditionary airfield units.[29] This deployment into CENTCOM's area of responsibility began in January 2003 and was accomplished with 670 airlifter sorties and sixty-two ship moves. During the course of the deployment, the RAF's four C-17s and other mobility aircraft transported roughly half of the personnel and equipment that needed to be moved by air.[30]

Australia's principal equipment contribution consisted of fourteen F/A-18 Hornet multirole fighters from 75 Squadron that deployed to Al Udeid Air Base via Diego Garcia. The Hornet's multirole capability made it a clearly

A stealthy USAF F-117 holds at the precontact position behind a USAF KC-135 tanker, with two USAF F-15Es at the inboard wing positions, two USAF F-16CJs at the middle wing positions, an RAF Tornado GR4 on the far right, and an RAAF F/A-18 on the far left. USAF photograph by Master Sergeant Ron Przysucha, at https://www.af.mil/News/Photos/igphoto/2000599821/.

preferred choice over the RAAF's only other combat aircraft alternative, its single-mission F-111C long-range maritime strike aircraft. In addition, the profusion of US Navy and Marine Corps F/A-18s that were also committed to the campaign offered the added advantage of commonality with respect to munitions, spare parts, and overall mission capability.[31]

All of the F/A-18s that the RAAF contributed were Hornet Upgrade (HUG) 2.1 jets that had recently been provided with improved APG-73 radars, APX-11 combined interrogator transponders, and an ASN-172 Global Positioning System (GPS) provision embedded within the aircraft's inertial navigation system. Because the upgrade program was still in an early phase at that time, HUG aircraft had to be marshaled from all three of the RAAF's combat-coded F/A-18 squadrons.[32] A select group of twenty-two experienced pilots was also gathered from the RAAF's three operational Hornet squadrons.[33] This detachment represented the largest deployment of RAAF fighters for combat since the Korean War. On arrival, 75 Squadron was embedded in the USAF's 379th Expeditionary Air Wing, which had also deployed to Al Udeid. In addition, the RAAF contributed two C-130Hs, one C-130J, and two AP-3Cs, each with appropriate combat-support personnel. Finally, the RAAF provided an air forward command element consisting of forty-two of its best air warfare experts to further augment CENTAF's overall CAOC staff of around 1,700 US and allied personnel at Prince Sultan Air Base. That liaison element was responsible for coordinating RAAF air operations with the other two coalition partners and providing national control of all RAAF assets that had been seconded to the coalition. Elements of an expeditionary combat-support squadron were also deployed forward to fulfill security, logistics support, airfield engineering, administrative, medical, communications, and other essential support functions.

TARGET-APPROVAL PROVISIONS

The British contingent was directly involved in CENTCOM's early selection of more than 900 target areas of potential interest throughout Iraq. The target-approval machinery for the British portion of OIF was developed in close consultation among the MoD, PJHQ, and the in-theater national contingent commander, who worked together to create the most expeditious approval arrangement possible. That machinery was developed and put into place to head off the sorts of target-approval delays involving coalition partners that had afflicted NATO's Operation Allied Force in 1999 and OEF two years later, when British government approval was required before CENTCOM could attack emerging targets with American aircraft that operated out of Diego Garcia or had just taken on fuel from RAF tankers. (The Australian target-approval process was very similar to the UK process.)

Target approval in the southern no-fly zone had proceeded at a relatively slow pace, and decisions about RAF attacks were routinely referred back to PJHQ and the MoD.[34] The anticipated rapid tempo of Operation Telic, however, would not allow that luxury. Target-approval decision making would have to go from "sedate" to "fast and furious," in Air Chief Marshal Burridge's formulation.[35] That requirement, in turn, dictated significant delegation of target-approval authority to Burridge and his targeting board. As was the case for the US side within CENTCOM's chain of command, powers delegated to the in-theater British national contingent included the authority to attack emerging targets quickly as actionable intelligence on those targets became available. Burridge later summed up his role in this respect succinctly: if a target were to be attacked with a British platform, either he or someone in the British contingent to whom he had delegated authority had to approve it. The same applied in the case of a US heavy bomber operating out of a British facility such as Diego Garcia or RAF Fairford.[36]

Target nominations that required approval at the highest levels were submitted through PJHQ to the MoD's targeting organization, whose principals would present the target requests to the appropriate ministers for approval. Such ministerial involvement with target approval was kept to an absolute minimum in the campaign. The secretary of state for defence laid out the broad parameters of what was acceptable and what was not and reserved beforehand certain target categories that only minister-level authority could approve; those categories, however, were very few. As one minister put it, "The military men were given maximum flexibility within those parameters to go about their task."[37] Torpy could not recall a single situation in which the British contingent's need to seek political clearance conflicted with immediate campaign operational requirements because so many contingencies had been anticipated and planned for in advance.[38] The only targets within the United Kingdom's cognizance that really needed London's approval were command-and-control targets that might result in a significant amount of collateral damage.

Consistent with the laws of armed conflict, target nominations stipulated that no target attack would be carried out if any anticipated loss of noncombatant life, injury, or other harm were deemed excessive in relation to the direct and material military advantage anticipated from the attack. Torpy was the designated "red card" holder should any suggested target over which the United Kingdom had veto power prove unacceptable for any reason. Furthermore, with strong British concurrence, CENTCOM planned effects-based operations and selected all targets with a view toward achieving a particular military effect.

British command elements did sometimes influence the selection of targets over which the United Kingdom did not wield veto power as well as of some targets that lay outside its formal purview altogether. In such cases, the American side freely accepted proffered British advice, even when not required to do so. Burridge later explained that the British contingent felt that it provided valuable input "in saying yes, okay, this is an American target, American platform, no British involvement, but actually let me just say how this might be viewed in Paris, Berlin, or wherever."[39] In its after-action report, the MoD noted that there had been no instances in which such proffered British advice had not been accepted. It further noted that the targeting authority that was delegated to British commanders was "significant" and allowed for flexible and responsive operations.[40]

RAAF and Australian special-operations forces legal officers were also assigned to the CAOC to ensure that targets assigned to 75 Squadron were "appropriate and lawful."[41] The Combined Targeting Coordination Board was chaired by Major General Victor Renuart, USAF, CENTCOM's director of operations, with the United Kingdom and Australia represented, respectively, by Burridge and Brigadier McNarn. Like Burridge, McNarn provided whatever senior military campaign advice to the board that he deemed useful and appropriate. ADF commanders also had service lawyers at their side who vetted targets on CENTCOM-developed strike lists that were assigned to 75 Squadron and assessed them in accordance with Australian legal obligations. Several target categories were subject to ministerial approval, and, as was the case with all other coalition aircrews, Australian pilots could and did abort attacks when there was concern for collateral damage or if the assigned target could not be identified and validated from the air.[42] In a belated but ultimately helpful additional contribution to this effort, the Australian Department of Defence announced on 31 March that it was sending six RAAF imagery-analyst officers and airmen to support USAF U-2 operations to help assess wet-film target images taken over Iraq, select targets, and assess strike results. In addition, a four-person RAAF battle-damage-assessment team was embedded within CENTCOM's headquarters at MacDill AFB.

The Australian air contingent commander, Group Captain Geoff Brown, reflecting on the instant trust relationship that he established with General Moseley on arriving in the CAOC, commented:

> I think we lucked in with the fact that he was in charge of CENTAF at the time, because my first day there . . . he took me under his wing and bang, sat me down at the table with him at the first [video teleconference] he was having. . . . It was interesting the way he treated me[;] . . . he let the rest of his colonels know that I had direct access to him. At the time, I didn't realize the value of that. . . . But the reality was that they were incredibly inclusive of us in the planning side[;] . . . they allowed us to have a look at the plan, where we wanted to go, how we wanted to operate, they pretty much allowed us to do what we wanted to do. . . . [General Moseley] was just happy to have us there.[43]

Punctuating this point, the commander of the RAAF's 75 Squadron, Wing Commander Melvin Hupfeld, later remarked in his own postcampaign reflections: "If you ever have to go to war, go to war with the Americans. They set things up very well."[44]

OVERALL COMBAT PERFORMANCE

The second Persian Gulf War was a genuinely combined operation when it came to British and Australian involvement, with RAF and RAAF sorties wholly integrated into the CAOC's daily air-tasking flow. For the RAF, these tasks included offensive strike, defensive counterair, surveillance and reconnaissance, tanking, intratheater lift, and aeromedical sorties. Offensive strike missions included firing a number of air-launched antiradiation missiles at the acquisition and tracking radars for Iraq's surface-to-air missiles.[45] In all, the RAF flew 2,519 sorties, approximately 6 percent of the coalition total of 41,000.[46] RAF tankers dispensed some 19 million pounds of fuel, more than 40 percent of which was transferred to US Navy and Marine Corps strike fighters.[47] (RAF tankers, with their drogue refueling system, are compatible with probe-equipped US Navy and Marine Corps aircraft, but not with many USAF tankers, which mainly employ the boom refueling system.)

One clear lesson the RAF had learned from its experience in Operation Allied Force was the need to improve its all-weather strike capability. After the Kosovo campaign, the RAF stepped out smartly to acquire antiarmor Maverick missiles and enhanced Paveway laser-guided bombs (LGBs) equipped with additional GPS guidance, thus building on its existing Paveway LGB capability. The RAF also increased the number of Tornado GR4s and Harrier GR7s that were capable of delivering such munitions.

The RAF in OIF operated out of eight forward locations in various countries throughout the region.[48] It employed fewer aircraft than it had in Desert Storm, but its greatly expanded use of precision munitions allowed those aircraft to produce substantially greater combat effects than before. RAF aircraft released 919 of the 29,200 munitions expended by the coalition, with roughly 85 percent being precision guided. Weapons released on RAF sorties included Storm Shadow, Paveway II, enhanced Paveway II, enhanced Paveway III, Maverick, air-launched antiradiation missiles, and unguided general-purpose bombs. The 138 unguided bombs included 70 cluster bombs that were used against enemy troops and armor in the open, primarily in the vicinity of Baghdad. The RAF also delivered a number of precision-guided inert 1,000-pound bombs in an effort to minimize collateral damage, but these concrete-filled shapes, which relied on kinetic energy, often did not produce the desired effect when they hit their designated aim point.

OIF saw the first combat use of the RAF's Storm Shadow precision stand-off ground-attack cruise missile, which was designed to penetrate and disable hardened structures. It was guided by GPS and digital terrain-profile matching, with a terminal seeker for maximum accuracy and collateral-damage avoidance, and was delivered by Tornado GR4s. It was used both day and night and in all weather conditions to attack a variety of high-value targets. The Storm Shadow missile was almost invariably accurate, offering what the MoD's initial after-action look called the promise of a "step change" in the RAF's precision standoff attack capability.[49] The missile had only just begun coming off the production line, so the RAF had a limited supply. It proved adequate, however, for the targets that Moseley designated. Indeed, apart from the B-2 stealth bomber armed with GBU-37 hard-structure munitions, Storm Shadow provided Moseley with the only significant deep-penetration target-attack capability available to CENTAF. Twenty-seven Storm Shadow missiles were fired during the course of the campaign, mostly during the first few days against especially hardened enemy command-and-control facilities.

Storm Shadow was able to disable four such key targets in the opening seconds of the air war. Its hard-target penetration features made it uniquely qualified to fill a critical niche.[50] Storm Shadow was attractive to CENTCOM's weaponeers because it offered a better hard-target penetration capability than anything in the American standoff munitions inventory.

The tactical-reconnaissance capability offered by the RAF's Tornados, Harriers, and Jaguars was likewise in short supply among US forces; only the US Navy's F-14s configured with the Tactical Air Reconnaissance Pod System offered a similar capability. Tornado GR4s equipped with Reconnaissance Airborne Pods for Tornado (RAPTOR) and Canberra PR9s provided high-quality, near-real-time imagery in the tactical-reconnaissance role. The Nimrod MR2, normally employed as a maritime surveillance platform, supported coalition operations in the Iraqi western desert, providing both surveillance and reconnaissance support and serving as an airborne radio relay platform. RAF Harriers and Tornados had thirty Thermal Imaging Airborne Laser Designator (TIALD) pods on hand and used some of them in a nontraditional way as a surveillance and reconnaissance asset for monitoring enemy tank positions and potential Scud launching sites.

The RAF had just completed a major midlife upgrade of its Tornados to GR4 standard, adding a wide-field-of-view head-up display, new avionics and forward-looking infrared systems, GPS navigation, cockpit-lighting modifications to allow the use of night-vision goggles, and additional modifications that enabled the aircraft to carry the new Storm Shadow missile and the RAPTORs, first employed in Operation Resinate. Operating the RAPTOR system required the Tornado pilot to fly straight and level for a significant period of time to record target images, which made for a predictable flight path. Accordingly, Tornado GR4s conducting reconnaissance missions were typically escorted by USAF Block 50 F-16CJs armed with High Speed Anti-Radiation Missiles.[51] Although successfully used in the close air support (CAS) role, the Tornado GR4 was not the best-suited aircraft for that mission because of its limited maneuvering performance at high altitude. Target-area searches often require a bank angle of thirty degrees or more at 20,000 feet, whereas the GR4 was designed for high-speed, low-level operations. Offsetting this liability in some mission profiles, however, was the GR4's range capability, which exceeded even that of the USAF's F-15E. A CAOC planner recalled that the GR4 "was the only aircraft that could make it up to Kirkuk on night one."[52]

In another important contribution, the RAF deployed four E-3D Sentries to provide continuous coverage of one of the four AWACS tracks that were constantly manned adjacent to and eventually over Iraq. In Operation Telic, they flew 127 missions from 12 March through 27 May 2003, with a 100 percent mission success rate.[53] Because the E-3D can refuel with both probe-and-drogue and boom systems, it was refueled by USAF and Air National Guard KC-135s and USAF KC-10s as well as by RAF Tristar and VC-10 tankers. (By the end of major combat, the six VC-10s deployed to Prince Sultan Air Base had flown 223 missions, with an average sortie length of 4.5 hours, no mission aborts, and a consequent operational success rate of 100 percent. One-quarter of the 3,700 tons of fuel they offloaded were transferred to US combat aircraft.[54])

Although the United Kingdom's unmanned aerial vehicle, Phoenix, offered less capability than the USAF's RQ-1 Predator, it nonetheless played a key role in supporting coalition land forces, primarily by geolocating ground targets. It was initially used around the clock. As the campaign progressed, however, it flew mostly at night to extract the greatest resolution from its thermal-imaging sensor. In all, Phoenixes flew 138 sorties, 23 ending with the platform being lost or damaged beyond repair and another 13 sustaining repairable damage. Most of the losses were due to technical difficulties associated with operating in Iraq's unforgiving spring weather. Helicopters from the UK Joint Helicopter Command, in addition, provided combat support to ground forces from land and sea bases. Royal Navy Lynx and Gazelle helicopters fired forty-nine tube-launched, optically tracked, wire-guided missiles in successful attacks against enemy tanks, armored personnel carriers, and bunkers.

In reflecting on the many reasons that accounted for the RAF's strong showing in the three-week campaign, Burridge noted that the technology gap between American and British air power had for years been quite small and was narrowed further by the RAF's experience at working almost in lockstep with US forces in enforcing the northern and southern no-fly zones over Iraq and in training together with American combat air forces in joint and combined exercises such as Red Flag.[55] An informed and thoughtful RAF observer remarks that the various challenges the British encountered during the lead-up to and execution of OIF were overcome by a combination of mutual dependence; good fortune (in everyone's having had the comparative luxury of a fairly unrushed planning period); both sides' ready willingness to engage in burden sharing; deep mutual trust at all levels, especially deep and strong interpersonal relations at the most senior levels; and "a motivation to find common ground and to engineer solutions to any problems that threatened the coalition's integrity."[56]

From the very start of CENTCOM's planning for OIF, the British contingent was brought into the division of labor between the two main coalition partners in a way aimed at leveraging British assets that might best complement and enhance American forces. "We very definitely were not there for the ride," Burridge recalled. "On the air side, we flew [only] seven percent of the sorties, but we provided a larger proportion of precision-guided munitions than did the Americans" as well as "niche capabilities . . . that the U.S. was lacking in, particularly tactical reconnaissance."[57] After the period of major combat ended, both Burridge and Torpy confirmed that in contrast to the experience of Operation Allied Force in 1999, final target-approval delegations in OIF had been far more flexible. An informed account notes that use of Britain's red card "was avoided, on more than one occasion, because the trust that existed at all levels of command allowed informal dialogue to preempt any potential formal action. This approach was absolutely pivotal in minimizing friction."[58]

In a related vein, Torpy later indicated that the British national contingent never felt that only RAF assets should support British land forces because "that would be an inefficient use of air power. Inevitably, we would not have sufficient [British] assets to provide cover, for instance, to a [British] land component 24 hours a day."[59] The use of air power, according to Torpy, has to be planned and prioritized centrally, with execution decentralized, for efficient employment of air resources.[60] At bottom, the British contingent benefitted from close commonality with its American counterparts in equipment, communications, and operational mindset as well as from close personal relationships starting at the four-star level and going all the way down to aircrews working conjointly and harmoniously at the tactical level.

Australia's contribution to the three-week air war began almost as soon as the RAAF's 75 Squadron arrived in theater and was declared operationally ready. The squadron sought to gain an early combat edge by joining the RAF in Operation Southern Watch so that Australian pilots could be exposed to the CAOC's mode of operations in managing the airspace over southern Iraq, but the Australian high command refused to grant permission. A US Marine Corps exchange pilot with the squadron later reported, "Although the coalition wanted Australians to fly [Southern Watch] missions, the federal government in Canberra had sent us [75 Squadron] . . . [only] to support 'offensive operations' against Iraq. In the government's eyes, [Southern Watch] didn't warrant our participation. As termed by the commander of Australian forces, that would have been 'mission creep.'"[61] In fairness to the Australian government's position in this regard, enforcement of the no-fly zone under the UN mandate was never one of the reasons why the ADF was deployed to the Persian Gulf region.[62]

The RAAF's Hornets were initially assigned to provide defensive counterair protection for such high-value coalition aircraft as the E-3 AWACS, E-8 Joint Surveillance Target Attack Radar System, RC-135 Rivet Joint, and allied tankers operating near and over southern Iraq, with 75 Squadron typically generating twelve sorties a day toward that end. These defensive counterair missions typically lasted between five and six hours, with the F/A-18s conducting three or four in-flight refueling evolutions from allied tankers in the process.

With respect to the Australian contingent's confidence level going into the RAAF's first shooting war since Vietnam, the commanding officer of 75 Squadron, Wing Commander Hupfeld, later reflected on the question candidly and at considerable length.

> The Americans had been fighting there for 12 years, so they knew what Iraq had. That was shared well with us as Australians, so we had a good idea of the threats, and we pretty much trained to what we thought the worst case would be. . . . Our basic level of training was right where we needed it, so it was quite a simple process to actually then focus it more on the threats that we expected in the Iraqi theatre. . . . I was very comfortable that our aircrews knew what was required, and there was a broad range of skills that we needed. . . . [Yet] I was very comfortable with our level of skill at that stage and comfortable with the capabilities of the aircraft.[63]

At the outset, when 75 Squadron's fighters were principally assigned to defensive counterair combat air patrols (CAPs), with the option to be assigned another role as necessary for strikes against emerging ground targets, the typical weapons loadout for each aircraft consisted of two wingtip-mounted

AIM-9M Sidewinder infrared air-to-air missiles, three AIM-120 advanced medium-range air-to-air missiles, and a single 500-pound GBU-12 LGB. Just a day into the campaign, however, the CAOC asked a 75 Squadron Hornet escorting a high-value coalition aircraft to strike a designated ground target that had just emerged and been validated. After confirming that the requested attack was consistent with Australian laws of armed conflict and standing rules of engagement, the Australian air contingent commander, Group Captain Brown, approved the attack. One of his F/A-18s promptly dropped the first Australian bomb released in anger since a 2 Squadron Canberra light bomber last did so during the Vietnam War (see Paget's chapter "Magpies and Eagles"). The entire request-and-approval sequence took less than thirty minutes, with an initial bomb-damage assessment provided to the ADF's forward headquarters ten minutes thereafter.[64] Within forty-eight hours of the campaign's commencement, the CAOC switched 75 Squadron to a "swing mission" configuration, swapping out one of the three advanced medium-range air-to-air missiles for an additional GBU-12 or 2,000-pound GBU-10.

On 24 March, the RAAF's chief, Air Marshal Angus Houston, reported that 75 Squadron's Hornets were being assigned strike missions against selected fixed targets in Iraq, even as they continued to fly defensive counterair sorties as required by the CAOC. In the strike role, RAAF fighters flew as autonomous formations within a larger coordinated package of coalition aircraft, with one such early strike featuring an RAAF Hornet pilot serving in the role of mission commander.[65] As the campaign moved closer to its endgame and the squadron found itself performing almost exclusively CAS missions, smaller 500-pound GBU-12s replaced the GBU-10s.

Brown recalled that

> in the initial planning, the USAF wanted to have about four CAPs [provided by 75 Squadron's fighters], but we were tanker-limited, so we ended up just having the three CAPs. . . . We normally had a counter-rotating CAP of two aircraft and [another] two aircraft would be on the tanker at any one particular time. Just to maintain that over an eight-hour period—which we did—took 12 of our 14 aircraft. So over [the first] nine days, we had to keep 12 of the 14 F/A-18s serviceable. That was a challenge, as the logistic system did not work quite as well as I would have liked.[66]

Nevertheless, the RAAF's F/A-18s managed to maintain a better than 90 percent fully mission-capable rate throughout the campaign. It was fortunate, Brown added, that the Australian government had deployed the Hornet detachment more than a month before the campaign got started because "it took us probably about two weeks to make sure we had communications links and everything working properly."[67]

Once it became apparent to CAOC planners after about nine days of combat that the Iraqi air force was likely to remain out of the fight, the RAAF's Hornets were swung almost exclusively to ground-attack operations. The first preplanned strike mission took place even before that determination when a four-ship flight of 75 Squadron F/A-18s was integrated into a strike package that also included American and British strike and electronic-warfare aircraft in an attack against Republican Guard units near Al Kut. In addition, as the allied land offensive reached full swing, RAAF Hornets contributed to kill-box interdiction/CAS operations in support of both US Army V Corps and I Marine Expeditionary Force [I MEF]. Because the RAAF's F/A-18s were not configured to carry GPS-aided Joint Direct Attack Munitions, they dropped either the GBU-10 2,000-pound LGB or the GBU-12 500-pound LGB, depending on target weapon requirements; these two precision munitions types accounted for all of the bombs delivered by RAAF fighters during the campaign.[68]

By 29 March, 75 Squadron's Hornets had flown defensive counterair, strike, and CAS missions, with CAS sorties typically being diverted from initially assigned defensive counterair missions. During the campaign's final days, when kinetic attacks became increasingly infrequent, the RAAF's Hornets would join other coalition strike fighters on request from ground commanders to provide so-called shows of force by making low-altitude, high-speed passes over concentrations of Iraqi civilians to break up gathering crowds that were causing problems for allied occupation forces. By the deployment's end, the fourteen RAAF Hornets had flown 1,800 hours and more than 670 sorties, with more than 350 of them combat sorties, and had dropped 122 LGBs on assigned enemy targets, all to useful combat effect.[69]

RAAF C-130s also supported allied operations in southern Iraq, with their first mission to airlift ground refueling trucks into the recently captured Tallil Air Base near An Nasiriyah on 30 March so that coalition aircraft could use the base as a forward-operating facility. Although the three C-130s provided by the RAAF represented only 3 percent of the total coalition Hercules force, they lifted 16 percent of the coalition's total cargo delivered by C-130s into CENTCOM's area of operations. By the time the Ba'ath regime fell, 36 Squadron's C-130Hs and 37 Squadron's C-130Js had exceeded 2 million pounds of cargo and more than 700 passengers delivered since the start of the deployment. Finally, the RAAF also operated one of its two deployed AP-3C Orions at various time windows over the North Arabian Gulf, using the aircraft's onboard sensors to detect and identify vessels in or near Iraqi waters, with special interest in any such vessels that might threaten coalition and civilian maritime operations by means of mine laying or suicide attacks. By 11 April, two days after the Ba'athist regime collapsed, the RAAF's Orions had maintained a near-perfect fully mission-capable rate of 98 percent.[70]

In a message issued five days before the fall of Baghdad, Air Marshal Houston quoted the chairman of the American Joint Chiefs of Staff, General Richard Myers, USAF, who had declared that "the contributions from the Australian force have been tremendous. . . . They have been absolutely superb, and we appreciate it."[71] The RAAF's air contingent commander, Group Captain Brown, was awarded the US Legion of Merit by General Moseley, as was the RAF's air contingent commander, Air Marshal Torpy.

Brown reported afterward that there had been no major interoperability issues with the F/A-18s, C-130s, or P-3s. He added:

> If you ask any of the guys who flew in the missions over Iraq, they were pretty happy with the [rules of engagement] that they had, and they were pretty happy with the targeting directives. I believe that that was a very mature approach by everybody who was involved in the chain. In the CAOC, I also had exactly the same sort of collateral damage criteria as [General Moseley did] and, again, that made it easier to operate [in a coalition context]. . . . The fact that we fitted into the operation seamlessly, I think, had a lot to do with the training regimes that we have had over the last 50 years.[72]

Among the operating issues that 75 Squadron encountered, Brown singled out the Nighthawk targeting pod (the same one carried by US Navy and Marine Corps F/A-18s before the advent of Advanced Targeting Forward-Looking Infrared) as "the one that gave me the most heartache over the five weeks" because it "was unreliable and . . . did not have enough magnification capability." He also mentioned the RAAF's lack of night-vision goggles, an inadequate electronic-warfare suite, and a lack of all-weather standoff precision munitions such as the satellite-aided Joint Direct Attack Munitions series.[73] As for the many positives, he concluded:

> The coalition with whom we worked, and the USAF in particular, were incredibly accommodating. The USAF allowed us to do pretty much what we asked them. I do not think I was ever knocked back on any particular request. . . . If you are going to work with anybody, you cannot pick a better partner. We also had the RAF there and, again, they were incredibly generous to us. I did not have a communications aircraft to get from where I was to the other bases, but the two-star there [Air Vice Marshal Torpy] gave me a pretty free rein in his HS 125. So great cooperation . . . really made the difference.[74]

One minor problem experienced by the RAAF's Hornet pilots and often, no doubt, by US Navy and Marine Corps and RAF pilots, whose aircraft were similarly configured with the probe-and-drogue in-flight refueling system, had to do with refueling from the USAF's KC-135 tanker, which normally uses the boom refueling system but also is equipped to provide probe-and-drogue refueling for non-USAF aircraft.[75]

Another reported source of frustration for 75 Squadron's pilots, one that was encountered at times by all coalition combat aircrews, entailed the inefficient apportionment of kill-box interdiction/CAS targets by the USAF Air Support Operations Center, which supported V Corps, due to the V Corps commander's insistence on close control of kill-box operations on his side of the fire-support coordination line. Like the other allied strike pilots, those in 75 Squadron soon eagerly sought assignment to the far more efficient Marine Corps Direct Air Support Center, which supported I MEF, so that they might have greater opportunities to employ their GBU-12s to useful effect rather than returning to Al Udeid with unexpended munitions. At first, the ADF's air contingent commander and the commander of Australian forces denied the transfer request because it smacked of "mission creep." Eventually, however, as the US Marine Corps' exchange pilot with 75 Squadron recalled, "after figuring out that Warhawk [the USAF Air Support Operations Center's call sign] was not a good organization to work for, the Aussies asked to get into I MEF's area of responsibility and were accepted—as any jet carrying bombs was."[76]

CONCLUSION

The coalition produced an effective combat performance throughout the air campaign during OIF. The air forces were able to build upon their past experience of conducting multinational exercises and operations. The nurturing of strong working relationships between key personnel underpinned the cohesiveness of the coalition. The early involvement of both Australia and the United Kingdom in the planning process cultivated trust and respect and provided an opportunity to influence operations and targeting. A clear target-approval process further provided a chance to shape operations. The involvement of the coalition air forces was, importantly, an added value as the availability of certain capabilities, such as Storm Shadow, provided great utility. The process of burden sharing in a range of areas, including refueling, added further value. The successful integration of the air forces was not without challenges, not least the strategic uncertainty caused by the doubts about the involvement of both countries in the forthcoming campaign, but they were overcome successfully. It was recognized, however, that the cohesion achieved would need to be sustained in the future.

Perhaps most important from an operational perspective was the serious and legitimate concern within the British defense establishment, particularly within the RAF, that the close commonality of operational styles and the trust relationships established between British and American airmen through their combined involvement in enforcing the no-fly zones over Iraq would disappear once Northern Watch and Southern Watch were no longer needed. In this regard, RAF officers from the Chief of the Air Staff on down acknowledged a pressing need to replace that former real-world marriage of forces with surrogate peacetime mutual-training opportunities that would be regularly exercised either in the United States or wherever else the airspace and required training infrastructure might allow.

To head off or ameliorate this mounting concern, the USAF and the RAF in 2005 implemented a new engagement initiative aimed at sustaining close ties between the two services to help ensure that their long-standing interoperability efforts and that joint training and dialogue would continue to flourish. An early testament to this continued commitment was the successful conduct of the first Coalition Flag exercise involving USAF, RAF, and RAAF aircrews at Nellis AFB in January 2006. More important yet, RAF and RAAF officers continued to serve in key positions in CENTCOM's CAOC at Al

Udeid Air Base after the period of major combat in OIF ended, with air commodores performing rotational duties both as British and Australian air contingent commanders and as CAOC director. RAF aircrews continued to contribute to the counterinsurgency air operations over Iraq and Afghanistan that ensued at varying levels of intensity in each country.[77]

NOTES

1. This chapter is an edited excerpt from Benjamin S. Lambeth, *The Unseen War: Allied Power and the Takedown of Saddam Hussein,* © 2013 RAND Corporation. Reprinted by permission from the Naval Institute Press, Annapolis, MD.

2. For more on that contribution by the United Kingdom, see Benjamin S. Lambeth, *Air Power against Terror: America's Conduct of Operation Enduring Freedom* (Santa Monica, CA: RAND Corporation, 2005), 116–19. See also Nora Bensahel, *The Counterterror Coalition: Cooperation with Europe, NATO, and the European Union* (Santa Monica, CA: RAND Corporation, 2003), 55–63.

3. As but one indicator that British involvement in the prospective war against Iraq was anything but assured at that point, RAF aircraft were proscribed from taking part in some elements of Operation Southern Focus (author's conversation with senior staff officers at Headquarters RAF Strike Command, RAF High Wycombe, UK, 28 October 2004).

4. John Keegan, *The Iraq War* (New York: Knopf, 2004), 125.

5. Air Chief Marshal Sir Brian Burridge, RAF, "Iraq 2003—Air Power Pointers for the Future," *Royal Air Force Air Power Review,* Autumn 2004, 7–8.

6. *Operations in Iraq: First Reflections* (London: UK Ministry of Defence, 2003), 3.

7. *Operation Telic: United Kingdom Military Operations in Iraq* (London: Report by the Comptroller and Auditor General, HC 60 Session 2003–2004, 11 December 2003), 1.

8. Ibid., 1.

9. David Willis, "Operation Iraqi Freedom," *International Air Power Review,* Summer 2003, 16–18.

10. In connection with Operation Slipper, the Australian government dispatched an Australian national commander to Kuwait in October 2001. It also concurrently deployed a task group of special-operations forces to Afghanistan, four F/A-18 Hornet fighters for the local air defense of Diego Garcia, and two Boeing 707 tankers to support allied air operations operating out of Manas, Kyrgyzstan, against the Taliban and Al Qaeda in Afghanistan. In early 2003, it sent two P-3 Orion maritime patrol aircraft to the Persian Gulf region.

11. The first RAAF officer, Wing Commander Otto Halupka, joined CENTAF's initial planning deliberations at Shaw AFB as early as August 2002, after having deployed shortly before for initial in-briefs at CENTCOM's headquarters at MacDill AFB with the designated Australian national commander, army brigadier Maurie McNarn, and the designated Australian air contingent commander, Group Captain Geoff Brown (official interview with Group Captain Geoff Brown, RAAF, Operation Falconer Air Component

Commander, 9 April 2008, provided to the author by the RAAF Air Power Development Centre [APDC], Canberra, Australia).

12. Tony Holmes, "RAAF Hornets at War," *Australian Aviation,* January–February 2006, 38.

13. The Honorable John Howard, address to the House of Representatives, Parliament House, Canberra, Australia, 18 March 2003.

14. Testimony recorded in *Lessons of Iraq: Third Report of Session 2003–04,* vol. 1 (London: House of Commons, Defence Committee, HC 635, 8 June 2004), 33.

15. The MoD assigned a three-star RAF representative, Air Marshal Jock Stirrup, to CENTCOM headquarters in Florida at the start of OEF. Stirrup was later replaced by a two-star successor as the Afghan air war ramped down, but the position was again upgraded to a three-star billet for OIF with the assignment of Air Marshal Burridge as national contingent commander.

16. Author's conversation with Air Chief Marshal Sir Brian Burridge, RAF, Commander in Chief, RAF Strike Command, on board a 32 Squadron HS 125 en route from RAF Northolt to RAF Lossiemouth, UK, 27 October 2004.

17. Testimony recorded in *Lessons of Iraq,* 1:53.

18. Testimony recorded in *Lessons of Iraq: Third Report of Session 2003–04,* vol. 2 (London: House of Commons, Defence Committee, HC 635, 8 June 2004), Ev 43.

19. Claire Bannon, Media Liaison Officer, "Op Bastille/Falconer Timeline," n.d., Government of Australia, Canberra.

20. *The War in Iraq: ADF Operations in the Middle East in 2003* (Canberra: Australian Department of Defence, 2004), 13.

21. Testimony recorded in *Lessons of Iraq,* 1:32.

22. Author's conversation with senior staff officers at Headquarters Strike Command, 28 October 2004.

23. "RAF Contribution to Operation Iraqi Freedom," briefing given to the author by 23 Squadron, RAF Waddington, UK, 29 October 2004.

24. Testimony recorded in *Lessons of Iraq,* 2:Ev 44.

25. *The War in Iraq,* 8.

26. The Honorable John Howard, address to the Committee for Economic Development of Australia, Canberra, 20 November 2002.

27. HMA ships *Darwin* and *ANZAC,* along with the RAAF's AP-3s, were already deployed in CENTCOM's area of responsibility in support of Operation Slipper, the ADF's contribution to OEF against the Taliban and Al Qaeda in Afghanistan.

28. Author's conversation with Air Vice Marshal Andy White, RAF, Air Officer Commanding, No. 3 Group, Headquarters RAF Strike Command, RAF High Wycombe, UK, 28 October 2004.

29. *Operations in Iraq: First Reflections,* 43–48.

30. Ibid., 43–48.

31. Holmes, "RAAF Hornets at War," 38.

32. Six aircraft came from 75 Squadron, and four each were drawn from 3 and 77 Squadrons.

33. Official interview with Group Captain Bill Henman, RAAF, Operation Falconer Commander, Air Combat Wing, 9 April 2007, provided to the author by the RAAF APDC.

34. Author's conversation with Air Marshal Glenn Torpy, RAF, Commander, PJHQ, Northwood, London, 26 October 2004.

35. Testimony recorded in *Lessons of Iraq*, 2:Ev 52.

36. *Lessons of Iraq*, 1:59.

37. Testimony recorded in ibid.

38. Torpy further volunteered that experience from previous operations had demonstrated the need for delegated targeting responsibility in order to maintain a sufficiently high tempo during Operation Telic. He also noted that the British defence minister, Geoffrey Hoon, fully understood those needs and was pivotal in securing the streamlining of the approval process (author's conversation with Air Marshal Torpy, 26 October 2004).

39. Testimony recorded in *Lessons of Iraq*, 1:59.

40. *Operations in Iraq: Lessons for the Future* (London: UK Ministry of Defence, December 2003), 27.

41. Bannon, "Op Bastille/Falconer Timeline."

42. On this point, ADF legal officers vetted targets "in concert with RAAF intelligence personnel to produce what was called a 'legal target appreciation' for each target to be hit by RAAF aircraft. Also, each target was briefed to Group Captain Brown for his approval. This was done in parallel with the US process, not sequentially, so as to not hold up the CAOC process. Anything outside Group Captain Brown's approval level was either not accepted as a target or, if there was time, referred back to Australia for a decision. If there was an issue with a target, we were always able to pull it from the US allocation to us and have it replaced with something more suitable. For this level of involvement, we relied heavily on RAAF planners within the GAT [guidance, apportionment, and targeting] and MAAP [master air attack plan] cells" (comments on an earlier draft by Group Captain Richard Keir, RAAF, Chief of Intelligence and Targeting for the RAAF under Group Captain Brown, Director, RAAF APDC, 2 July 2009).

43. Official interview with Group Captain Brown, 9 April 2008.

44. Official interview with Wing Commander Melvin Hupfeld, RAAF, Operation Falconer Commanding Officer, 75 Squadron, 16 September 2003, provided to the author by the RAAF APDC.

45. Author's conversations with senior staff officers at Headquarters RAF Strike Command, 28 October 2004.

46. *Operation Telic*, 7.

47. *Operations in Iraq: First Reflections*, 14.

48. *Lessons of Iraq*, 2:Ev 200.

49. *Operations in Iraq: First Reflections*, 23. The munition includes an initial penetrator charge followed by the main explosive charge in order to breach reinforced structures (Neil Baumgardner, "RAF Spokesman: Enhanced Paveway 'Outstanding' in Operation Iraqi Freedom," *Defense Daily*, April 2, 2003).

50. The loss of the use of Turkish bases for conducting combat operations into Iraq created a major ripple effect as the RAF's Tornado GR4 Storm Shadow shooters were forced to strike critical hard targets in northern Iraq that had originally been scheduled to be attacked by F-15Es configured with GBU-28 5,000-pound bunker busters. The GR4 was able to range farther north into Iraq than was the F-15E from the latter's southern

base at Al Jaber, Kuwait (author's conversations with Colonel Douglas Erlenbusch, Major Anthony Roberson, and other CENTAF staff, USAF, Shaw AFB, SC, 29 January 2007).

51. Flight Lieutenant Andy Wright, RAF, "GR4 Reconnaissance: Operation Telic," in *RAF 2004* (London: UK Ministry of Defence, Directorate of Corporate Communication, RAF, 2004), 24–25.

52. Comments on an earlier draft of this chapter by Lieutenant Colonel Mark Cline, USAF, 11 January 2008.

53. "RAF Contribution to Operation Iraqi Freedom."

54. Squadron Leader Hugh Davis, "VC-10 Tanker: Operation Telic," in *RAF 2004,* 44, 49.

55. *Lessons of Iraq,* 2:Ev 47.

56. Sophy Gardner, "Operation Iraqi Freedom: Coalition Operations," *Air and Space Power Journal* 18, no. 4 (Winter 2004): 98.

57. Testimony recorded in *Lessons of Iraq,* 2:Ev 47. This niche expertise in tactical reconnaissance was valuable because the USAF had withdrawn its dedicated RF-4C tactical-reconnaissance aircraft from service after the conclusion of Operation Desert Storm in 1991 (Michael Gething, Mark Hewish, and Joris Hanssen Lok, "New Pods Aid Air Reconnaissance," *Jane's International Defence Review,* October 2003, 52).

58. Gardner, "Operation Iraqi Freedom," 92.

59. Testimony recorded in *Lessons of Iraq,* 1:62.

60. Author's conversation with Air Marshal Torpy, 26 October 2004.

61. Holmes, "RAAF Hornets at War," 39. Nevertheless, as the commanding officer of 75 Squadron later recounted, the RAAF's deployed Hornet pilots did get some useful in-theater acclimation short of participating in Southern Focus before the campaign formally kicked off: "We did some missions where we flew up the North Arabian Gulf along the administrative route that would take us into the Iraqi theater. We flew up over Kuwait as well and did some training over there. While they were massing the army forces in Kuwait, we used that for our own training benefit to look at what a tank would look like on the ground, how we could see it on the radar or our other sensors, and so we experienced the communications process, the operation and coordination with our command and control agencies. We did all that and practiced that prior to actually deploying across the line, so we became quite familiar with what was there. We just hadn't experienced going across the border into a combat area" (official interview with Wing Commander Hupfeld, 16 September 2003).

62. Comments on an earlier draft by Air Vice Marshal Geoff Brown, RAAF, Deputy Chief of Air Force, RAAF, 2 July 2009.

63. Official interview with Wing Commander Hupfeld, 16 September 2003.

64. In his later recollection of this precedent-setting event for the Australian defense establishment, the RAAF's chief targeting adviser to CENTCOM characterized the final decision-making process: "The target was an Iraqi leadership target. . . . It just so happened that at the particular time that the target was detected and went through the approval chain, our aircraft were the closest ones to it. So after a quick national validation, I remember standing next to Geoff Brown with the lawyer in the CAOC, and I can't remember the exact words, but it was really: 'Are you happy, are you happy, are you happy' sort of thing, and we sort of said: 'Yep, it all appears good to us, your decision now, Boss.' So he said, 'Do

it'" (official interview with Wing Commander Richard Keir, 19 February 2007, provided to the author by the RAAF APDC).

65. Air Marshal A. G. Houston, RAAF, "Message from Chief of Air Force: Operation Falconer," Canberra, Australia, 24 March 2003.

66. Air Commodore Geoff Brown, RAAF, "Iraq: Operations Bastille and Falconer—2003," in *Air Expeditionary Operations from World War II until Today*, ed. Keith Brent (Canberra, Australia: Air Power Development Centre, 2008), 129.

67. Ibid., 127–28.

68. *The War in Iraq*, 26–28.

69. Ibid., 29.

70. Bannon, "Op Bastille/Falconer Timeline."

71. Air Marshal A. G. Houston, RAAF, "Message from Chief of Air Force: Operation Falconer," Canberra, Australia, 4 April 2003.

72. Brown, "Iraq," 131.

73. Ibid., 132–33.

74. Ibid., 132.

75. Holmes, "RAAF Hornets at War," 39.

76. Ibid., 42.

77. Author's conversation with Air Chief Marshal Sir Glenn Torpy, RAF, Chief of the Air Staff, London, 19 May 2009.

United Nations Peace Operations

*Precedents and Progress in the
Four Facets of Air Power*

A. Walter Dorn

When one thinks about United Nations (UN) peacekeeping, the first image that comes to mind is usually of a soldier on the ground wearing a blue helmet, possibly looking through binoculars to observe opposing forces or delivering humanitarian aid or patrolling in an armored vehicle to reassure the local population.[1] However, the peacekeepers of the air also play a vital role in UN operations, often serving similar functions. They deliver essential humanitarian supplies to war-affected populations. They observe from above to complement the ground view. Their aircraft can also send powerful signals by visible presence or by radio, light, or loudspeaker (from helicopters) that can act as a deterrent to wrong-doers and as a reassurance to locals and UN ground forces. When the local population or peacekeepers on the ground are under attack, UN aircraft can use their weapons to stop aggressors in their tracks.

Air power brings four facets (core capabilities) to military operations in general and to peace operations in particular: transportation, observation, communication, and firepower—in simple terms: carrying, seeing, signaling, and shooting.[2] The UN has utilized all four facets during its peacekeeping history, though it has sparingly used combat power, given the nature of these operations. This chapter examines cases in UN history from the 1960s to the present day to illustrate these four facets. It shows both the achievements of and the challenges faced by the world organization in utilizing air power to maintain international peace and security.

To support its noble goal, the UN has on occasion gone beyond peace-keeping to authorize enforcement operations in order to repel or reverse aggression—for example, in Korea in 1950 (see Williamson's chapter "The Korean War") and in the Kuwait–Iraq border area in 1991 (see Hallion's chapter "Coalition Air Power in the Gulf War, 1991"). But rather than use peace operations run by the UN secretary-general, these high-intensity combat operations are carried out by coalitions of states, usually with US leadership. In Korea, the UN Security Council authorized the United States to take the lead of "UN Command," a force of seventeen nations that reversed the North Korean invasion of the Republic of Korea. That mission used the most sophisticated aircraft of the day, including planes from aircraft carriers, and saw aerial dogfights with Chinese aircraft.[3] UN peace operations, by contrast, are established primarily to keep, foster, and build peace, so they rarely use combat. But there are exceptions.

One early example of a mission using all four facets of air power, including air combat, was in the Congo in the first half of the 1960s. Though unusual for a UN peace operation during the Cold War, the UN Operation in the Congo (ONUC) provides an excellent case to examine the applications of air power to robust UN operations. It was a forerunner of modern multidimensional operations, which have become prevalent in the twenty-first century.

THE CONGO OPERATION, 1960–1964

Dubbed the UN's first "air force," the air fleet in the service of ONUC was remarkable.[4] The United States initially provided the air transportation, using more than fifty C-124 Globemaster II and C-130 Hercules aircraft to bring most of the 20,000 peacekeepers from twenty-eight countries to the fractious Congo.[5] Canada then led in logistical sustainment flights with regular DC-4 North Star flights between Pisa, Italy, and the Congo's capital, Leopoldville (current-day Kinshasa).[6] Sweden provided J-29 Tunnan jets, some specially equipped for photoreconnaissance, as well as air intelligence officers to produce actionable intelligence. Other J-29 jets were used in combat, firing their rockets and machine guns on secessionist Katangan forces on several occasions. India also provided aircraft that conducted reconnaissance and applied armed force with B(I)58 Canberra light bombers.[7]

Since ONUC was a peacekeeping operation, the United Nations was initially reluctant to bring combat aircraft to the Congo. But lacking this facet of air power had significant and unwanted consequences, which became all too apparent when three CM-170 Fouga Magister trainer aircraft from the secessionist Katanga Province attacked the UN's operation, including air supply flights. Fortunately for the UN, one of the original Fougas was lost when its pilot tried to fly under a power line. UN ground forces managed to capture another on a secessionist-held airfield. But the remaining plane gained world attention during the hostilities of September 1961. The single jet, flown by a Belgian mercenary, strafed UN positions, including UN Headquarters in Katanga, and isolated a company of Irish troops operating under the UN flag, who were forced to surrender to Katangan forces. Furthermore, the Fouga jet destroyed several UN-chartered aircraft at Katangan airfields. A US State Department official, Wayne Fredericks, wryly commented: "I have always believed in air power, but I never thought I'd see the day when one plane would stop the United States and the whole United Nations."[8]

In the late summer of 1961, UN secretary-general Dag Hammarskjöld made an attempt to negotiate with the secessionist leader of the mineral-rich Katanga Province, Moïse Tshombé, who was strongly backed by Belgian mining interests that funded his mercenaries. Having grown increasingly concerned about the threat from the secessionist air force, the UN-contracted Swedish DC-6B carrying Hammarskjöld maintained mostly radio silence and flew at night. The previous day, the plane had been targeted by ground fire. Then on 18 September 1961 the plane crashed under suspicious circumstances on its approach to Ndola Airport in Northern Rhodesia, where the negotiations were to take place. With Hammarskjöld's death, the international community mobilized. The battle for Katanga entered a new phase, and the UN realized it needed both aerial reconnaissance and combat air power.[9]

In an attempt to obtain intelligence on Katanga's air capability, ONUC began mandatory debriefings of its transport aircrews after landing, having noted that they were making valuable observations during their flights and stops in the Congo. The mission also created an air reconnaissance unit. In 1962, Sweden provided two J-29Cs, the photoreconnaissance versions of the J-29 Tunnan jet, the "Flying Barrel." Also, Indian B(I)58 Canberra light bombers had their bomb bays outfitted with high-resolution cameras.

With the observation facet of air power being integrated into its airborne capabilities, ONUC now had actionable intelligence from airborne sources to improve its operations. Indian Canberras were soon dispatched to attack Katangan airfields. The UN action created havoc among Katangan forces in much the same way that the armed Fouga had earlier done to ONUC. This response prompted an aerial arms race between the UN and Katanga. By 1 October 1962, the UN mission estimated that the Force aérienne katangaise (FAK) had twelve Harvard single-propeller aircraft, eight or nine Fouga trainer jets, four T11 Vampire jet fighters, and a large number of P-51 Mustang single-propeller fighters being delivered to it. By contrast, the UN mission possessed six Canberra jet fighter-bombers, four Saab J-29B fighter-bombers, and four F-86 Sabre jet fighters. Furthermore, India soon had to withdraw its Canberras because of its war with China, which began later that month.[10]

The United Nations deliberately decided not to destroy FAK aircraft on the airfields for fear that FAK would then retaliate on UN airfields and because of a concern that FAK would disperse its aircraft to other airfields, thereby making the task of locating and destroying these aircraft much more difficult. The ONUC commander instead decided to wait until ONUC possessed a fighter force capable of destroying the bulk of Katanga's air force on the ground in an overwhelming surprise attack.[11]

The ONUC commander's "wait until ready" strategy proved to be wise. Near the end of 1962, with more combat aircraft in the "UN Air Force," ONUC was poised to strike jointly with UN ground troops under a plan appropriately named "Grand Slam." Katangan forces conveniently provided the justification for the operation when on Christmas Eve Katangese gendarmes shot at a UN observation helicopter, fatally wounding an Indian crewmember and, after the aircraft was forced down, seized and beat the crew. Elsewhere, Katangan forces began firing continuously at UN positions, fatally wounding several soldiers. After four days of cease-fire initiatives, the UN commander in Elisabethville, Major-General Prem Chand of India, finally persuaded UN Headquarters to approve a decisive offensive operation. The convincing argument came from radio intercepts revealing that Katanga's commander had ordered his air force to bomb Elisabethville airport during the night of 29 December.[12]

In ONUC's Operation Grand Slam, the mission's J-29 jet fighters, equipped with 20-millimeter cannons and 13.5-millimeter rockets, struck Katangan airfields and air assets with confidence, achieving a high level of surprise. Active patrolling of the skies by the Swedish J-29s effectively cut the air bridge between Katanga and its allies in Portuguese West Africa and Southern Africa, precluding the introduction of new aircraft. From 28 December 1962 to 4 January 1963, UN aircraft conducted a total of seventy-six sorties against Katanga's airfields, destroying most of the rebel aircraft on the ground.

During Operation Grand Slam, seven UN fighter aircraft and one reconnaissance plane were hit by ground fire, but none crashed, and no crew members were injured. In addition to kinetic action against Katangan air assets, UN fighter aircraft also provided close air support to UN ground forces. Katangan forces finally capitulated on 15 January 1963, and the province renounced its bid for secession.

An ONUC intelligence team subsequently learned that the Katangan air force still had some fifteen aircraft hidden at Angolan airfields. According to Belgian mercenaries, interrogated by the UN intelligence team, these aircraft were part of a planned buildup. One mercenary said to the UN intelligence team: "If you had only given us four more weeks so that we could have got the Mustangs ready, you would have experienced the same disastrous surprise one early morning at your Kamina [UN] Base as we experienced at Kengere [Kolwezi, FAK base] on 29 December."[13]

Through a combination of luck and military prowess, especially in the air, the UN prevented the secession of Katanga Province. However, difficult lessons for UN air power were obtained in the Congo mission regarding all four facets of air power. Transport flights to and within the enormous country—the size of western Europe—proved extremely difficult, meaning some ground forces were left without timely reinforcements. It was only with the urgent assistance of US and Canadian flights that the troops and supplies could be moved.

Reconnaissance flights based simply on pilot vision proved at times to be highly flawed, resulting in cases of completely erroneous assessments of conditions on the ground. In one instance, the pilot could not determine that peacekeepers on the ground were under siege and mistakenly interpreted the waving flags as a positive signal that things were okay! In another case, the pilot spotted Congolese running behind UN peacekeepers and, thinking the UN soldiers were being chased or hunted, considered firing on them. In fact, they were friendly Congolese who had saved the soldiers and simply wanted to help them get on their flights. In both incidents, the UN force managed to overcome the obstacles, but good ground-to-air communication was sadly lacking. The UN did use air power to communicate to the local population by dropping leaflets, especially before combat operations, to inform them that the UN was a force of peace.

Combat operations proved most difficult and controversial with the untested UN command-and-control system. At least two major combat operations, Operations Morthor and Rumpuch, were launched in Katanga in the fall of 1961 without proper approval, including approval by the UN secretary-general. They caused hundreds of casualties and threatened the very survival of the mission. The UN also had to deal with an emboldened "Katanga lobby" in New York and in world public relations. The lobby, funded by mining interests, argued against any use of force against the anti-Communist Katangan government. For other, ethical reasons, some UN pilots refused to use their weapons, even as close air support for ground troops under attack or to deal with menacing Katangan aircraft. In some cases, the Swedish government argued that its aircraft were to be used for defensive purposes only. In December 1962, the Swedish air contingent, to its credit, applied force robustly and successfully in Operation Grand Slam. This contribution was decisive in maintaining the Congo's territorial integrity.

The mission put considerable financial strain on the fledgling world organization, bringing it to the brink of bankruptcy, and resulted in much unwanted negative publicity and adversity, particularly from the well-organized backers of Katangan secession. The UN even faced allegations that its aircraft had inadvertently bombed a hospital and a hotel. The UN did not deploy any further peacekeeping operations in Africa until the end of the Cold War. The Congo operation was an unusual foray into multidimensional peace operations that was not characteristic of Cold War peacekeeping, which typically involved only observer missions and ground forces interposed between opposing armies. But ONUC did provide valuable lessons on what the United Nations could and could not do in using the resources of its member states. However, it was only in the twenty-first century that multidimensional peacekeeping with a strong aerial dimension came into regular usage.

Twenty-First-Century Peace Operations

The United Nations initially had a difficult time adjusting its peacekeeping practice after the Cold War. Many UN officials had forgotten the experience of the Congo. And the nature of peacekeeping shifted from the traditional model, where peacekeepers freeze the military side of a conflict so that a negotiated peace can be settled upon, to the new multidimensional peace-operations model that added the functions of peacebuilding (long-term infrastructure for peace), humanitarian assistance, and peace enforcement (when a party violates an agreement or commits atrocities). In all of these newer functions, air power plays a key role, and the UN had to (re)learn to incorporate aircraft into these new functions.

As in traditional missions, the transportation facet of air power has continued to play a key role in the new century, but the area of responsibility for new missions is now usually much greater, moving from confined border areas to entire nations. Widening the mission areas also brings added value from the observation facet of air power, complementing ground patrols to increase situational awareness. Unmanned aerial vehicles (UAVs) were first added in 2013 to complement manned aircraft. Aircraft are now used to relay signals and to communicate to both UN ground personnel and the local population. Most dramatically, combat aircraft, especially attack helicopters, are used in some missions. Each of these facets is worth examining in detail using select case studies.

TRANSPORTING AID: POSTEARTHQUAKE HAITI

For many, the first image of the modern UN air component is one of humanitarian aid delivery: food and medical assistance to those who have suffered through armed conflict or natural disaster. A prominent case is the Haitian earthquake of 12 January 2010. The half-island state of Haiti was struck by a catastrophic earthquake (7.0 magnitude), which was followed within two weeks by fifty-two aftershocks measuring 4.5 or higher on the Richter scale. Haiti, the poorest country in the Western Hemisphere, faced an immediate and overwhelming humanitarian crisis, with an estimated 200,000 killed by the quake. The state, already poor, suffered greatly and lacked the capacity to rescue its citizens from under the rubble of their own homes and workplaces or to sustain the displaced population.

The UN peacekeeping operation, the Mission des Nations unies pour la stabilisation en Haïti, which had been on the ground in Haiti since 2004, also suffered tragically from the quake. The main headquarters building and other UN facilities were crushed, resulting in the largest loss of UN personnel's lives in a single day in UN history (101 UN staff and family members, including the head of mission and the UN police commissioner). As the mission struggled to get back on its feet, the international community rushed in with aid.

The humanitarian mission in Haiti necessitated a high-intensity, restricted-infrastructure airlift operation. The main seaport of Port-au-Prince was inoperative due to the destruction, and so air transport took on central importance. A coordinated international effort was needed. To prevent the main airport in the Haitian capital, Port-au-Prince, from becoming gridlocked, the Haitian government turned over control of it to the US Air Force (USAF). However, some officials, including some from France and Italy, criticized the USAF in strong terms, calling it an "occupation" authority in Haiti and decrying its refusal to provide landing rights to some relief agencies. As criticism and pressure grew from dozens of governments and relief agencies, the USAF invited the UN Humanitarian Air Services (UNHAS) to help supervise air relief efforts.[14]

Although UNHAS is operated by the World Food Programme, it serves many UN operations, agencies, funds, and programs as well as nongovernmental organizations (NGOs) on a cost-recovery basis. As with UN peacekeeping, UNHAS does not own any of the aircraft it operates; rather, it contracts them through an online bidding competition from a pool of shortlisted contractors. UNHAS contracts more than fifty aircraft and has provided humanitarian relief in dozens of conflict zones and after natural disasters. As in Haiti, UNHAS often launches operations on short notice in far from ideal conditions, including landing and taking off from substandard runways, and in degraded facilities. In some locations, it has to operate without local air traffic control or proper weather forecasts and sometimes in a hostile environment--though this was not the case in Haiti.[15]

In Haiti, the USAF needed UNHAS to help with coordination. UNHAS staff arrived at Tyndall Air Force Base, Florida, on 24 January. Although the crisis had already peaked, there was plenty to do to regularize humanitarian aid delivery. UNHAS and the USAF worked together and honed their ability to cooperate in the massive humanitarian relief operation. Time slots for aircraft to land, unload, and disembark were given on a first come, first served basis, and the influx of air traffic meant that a significant air traffic backup had to be alleviated. UNHAS helped coordinate the air traffic and so abated the allegations that the United States was serving primarily US aircraft and interests first.

Although there were logistical issues with the time slots available for aircraft to land, unload, and disembark, the USAF and UNHAS operated effectively together. Although USAF personnel voiced negative perceptions of NGOs over matters such as arrival priorities and distribution procedures, the interaction between the USAF and UNHAS remained positive. UNHAS adjudicated between various competing NGOs that wished priority treatment. Neither USAF nor UNHAS had absolute control over Haitian airspace, so the government's sovereignty and authority were respected. Tensions sometimes arose when the government overrode or ignored the USAF–UNHAS time-slot procedures, but in general the overall coordination of the airlift relief operation was successfully done under USAF–UNHAS supervision and control.[16]

If transport is the first facet of air power and the most common, then observation is the second most common. The most significant development in twenty-first-century UN air power was the incorporation of unmanned aircraft.

Eye in the Sky: Unmanned Peace

The UN witnessed the utility of an unmanned aerial system (UAS, composed of one or more UAVs, a control station, and communications links) in the early 1990s as the United States deployed unarmed Predator drones in Bosnia over the heads of peacekeepers in the UN Protection Force. In the follow-up stabilization missions, starting at the end of 1995, the United States and the North Atlantic Treaty Organization (NATO) used UAVs extensively.[17]

In the Democratic Republic of the Congo (DRC) in 2006, the European Union Force deployed B-Hunter UAVs to support the UN during the first presidential elections in that country in decades. The UN subsequently began considering UAVs for its own operations. In 2007, the UN Observer Mission in Georgia (Eurasia) requested UAVs for surveillance, especially for the Kidori Valley in the breakaway region of Abkhazia. However, after a Georgian Hermes 450 UAV was shot down by a Russian MiG-29 in the spring of 2008, and after the Russo-Georgian War later that year, the Observer Mission in Georgia's mandate was not renewed due to a Russian veto at the Security Council.[18]

After two decades of toying with the idea and after two failed bidding processes, the UN finally contracted its first UAVs for peacekeeping in 2013.

At first, to make the point that its UAVs were not to be used for targeting and firing (in contrast to US Predator and Reaper drones), the UN called them "unarmed, unmanned aerial vehicles," or UUAVs. The UN contracted SELEX ES (currently under Leonardo S.p.A.) to operate five Falco drones for the UN Organization Stabilization Mission in the DRC (MONUSCO).[19] The introduction of these drones in the DRC quickly bore success.

On 7 February 2014, a Falco UAV provided useful imagery that showed a skirmish between a militia group (Alliance of Patriots for a Free and Sovereign Congo) and the Armed Forces of the DRC. On 26 April 2014, one UAV recorded the heat signatures of ten to fifteen people seen smuggling on the DRC-Rwandan border. The following month, a UAV on a training exercise spotted a sinking boat on Lake Kivu, thus allowing peacekeepers on zodiacs to save fourteen people.[20]

The introduction of UAVs in the DRC was initially condemned by Rwanda, which had a history of intervention and mineral exploitation in the eastern Congo. A Rwandan diplomat at the UN, Oliver Nduhungirehe, voiced concerns that the drones might gather intelligence on the entire region and not be limited to MONUSCO's area of responsibility.[21] Others also voiced concerns about drones being used for intelligence-gathering purposes. However, Rwanda's position changed quickly after the country's president, Paul Kagame, supported the UAV deployment and denied any accusations that Rwanda was aiding rebel groups, in particular the notorious M23.[22] UN rules stated that in order to fly UAVs near the Rwandan border, direct authorization by the head of mission was required. Some of the UN's troop-contributing countries had concerns, so far unfounded, that UAV footage could potentially be used to implicate UN troops in wrongful actions (such as criminal activity) or inaction, which would also erode the mission's principles.[23] Also, some humanitarian groups wanted to make sure that UAV usage would respect the principle of "humanitarian space," so that there was a clear physical separation between humanitarian and military aircraft in order not to compromise humanitarian neutrality.[24]

The Falco system had early problems. Its camera equipment initially had poor resolution; however, that difficulty was amended with the installation of an MX-10 camera.[25] The system also had ongoing technical problems, so that it was sometimes reduced to only one aircraft when four should have been available. But it proved the value of UAVs. These aircraft could remain airborne longer and so offered much longer loitering times than traditional

Hermes 900 UAV in its hangar in Timbuktu, Mali. The UAV was provided as part of a contracted service from Thales UK. Photograph by A. Walter Dorn.

aircraft, especially helicopters. The latter are typically limited to only a few hours of flight—often only two. Protracted loitering is a considerable benefit for surveillance. In addition, drones offer the advantage of reduced risk to operating personnel: they can be operated remotely and safely day or night or fly over regions where it has become "too hot" for peacekeepers due to hostile forces and fighting.

The UN expanded its use of drones considerably after the initial experience in the DRC. Medium-altitude, long-endurance systems, with control and images sent via satellite, were deployed in 2016 in the Multidimensional Integrated Stabilization Mission in Mali. Germany offered the Mali mission a Heron-1 medium-altitude, long-endurance UAV capability, based in Gao, while the UN contracted Thales UK to run a Hermes 900 system, based in Timbuktu.[26] The Hermes 900 UAV is shown in figure 12.1. These UAVs were often used for overwatch during ground patrols and in some cases were able to successfully warn UN convoys of improvised explosive devices being buried in the road in advance of their arrival.

The Swedish Intelligence, Surveillance, and Reconnaissance Task Force also deployed its own UAV systems in Mali. The task force operated the Textron Shadow 200 tactical UAV for surveillance missions in the Timbuktu region. The Shadow 200 had a mission endurance of nine hours and could operate up to 77.7 miles (125 kilometers) from its manned control station. Swedish units were also deployed with AeroVironment RQ-20 Pumas, mini-UAVs that are hand-launched and have a range of 9.3 miles (15 kilometers), which is especially useful for short-range reconnaissance in advance of motorized patrols.[27]

Micro-UAVs

Some drones are so compact in design that they are both man portable and deployable and can provide tactical surveillance for use by individual peace-keepers. The UN Multidimensional Integrated Stabilization Mission in the Central African Republic (MINUSCA) pioneered the UN's use of micro-UAVs with the DJI Phantom 4, a quadcopter weighing only 3.3 pounds (1.5 kilograms). Within one year of deployment of micro-UAVs in 2016, MINUSCA demonstrated more than sixteen innovative applications in four categories:

Engineering: Camp-change deduction and virtual-reality (three-dimensional) modeling; site-mapping survey; terrain visualizing; 360-degree line-of-sight visual survey

Peace Sustainment: Postconflict impact assessment; reconnaissance

Protection of civilians and **humanitarian action**: Monitoring of internally displaced persons (IDPs); protection of civilians during crises

Protection of UN mission (civilian, military, political): Camp-lighting assessment; camp-perimeter surveillance; camp water-stagnation areas; convoy-clearance assessment; crowd assessment; the planning of rescue/evacuation operations; VIP and other visit security

One example of the successful use of UAVs was in Kaga-Bandoro on 12 October 2016. When armed groups started fighting each other and attacking civilians, a micro-UAV was flown to observe the escalating conflict, proving especially useful since UN ground vehicles were not able to move freely. The UAVs helped to evaluate the situation and the extent of the ongoing damage; verify if humanitarian agencies required assistance, especially in case of evacuation; and identify the locations of IDPs. The UAVs could also observe unlawful movements and actions by armed groups and fires burning in different parts of the town. Seeing the conflict in real time from UAVs, the regional office head could pressure hesitant UN troops to go to the clearly identified dangerous areas to protect civilians. The UAS also spotted the movement of hundreds of IDPs toward the MINUSCA helicopter hangar, which prompted preventive action: Pakistani forces moved forward to protect the hangar as well as the IDPs, whose camp had been torched. With such positive UAV experiences, MINUSCA ordered forty additional DJI Phantom 4 drones for use by UN military observers in all twenty-one sites. Week-long training courses in micro-UAV operations were instituted for such observers and other staff.

Troop-contributing countries may now bring their own micro-UAVs to peace operations as contingent-owned equipment, yet they remain reluctant to deploy larger UAVs and the specialized personnel for image analysis due to the small number of such personnel available (even after the end of major operations in Afghanistan and Iraq). Although UAVs of all sizes are being adopted and integrated into many of the world's militaries, the technology is still sensitive due to the innovations that countries are undertaking. Because of this, many countries are reluctant to provide sensitive technology or reveal how advanced (or vulnerable) their technology is.

As the UN gains abundant imagery from UAVs and other camera feeds, however, there is a growing gap between the imagery acquired and the ability to analyze it. Finding videos of a given location taken several months or years earlier is proving difficult enough, let alone using cutting-edge applications on the imagery, such as pattern recognition, change detection, and other types of analyses that employ artificial intelligence and cloud computing. Actionable intelligence remains a key deficiency for UN missions.

The proliferation of commercial drone technology, especially micro- and mini-UAVs, means that the conflicting parties themselves are also increasingly using drones. Israel routinely violates Lebanese air space with UAVs, which the UN tracks with counterbattery radar from France based on the ground and radar onboard ships in the UN's Maritime Task Force off the Lebanese coast. Similarly, Hezbollah has flown UAVs into Israel, but they are comparatively primitive UAVs. Insurgents in Mali have flown UAVs over UN bases to look for vulnerabilities, though fortunately they have not yet used UAVs to carry munitions. In Libya, where a UN Special Political Mission struggles to mediate a civil war, the conflicting parties launch hundreds of drone strikes a month in what has been called the "the first war fought primarily by clashing fleets of armed drones."[28] The UN mission may need its own (unarmed) drones to monitor these battles. In this mission and others, it definitely needs counter-drone technologies, including radar for UAS detection. And to deal with threats of any kind, whether unmanned or manned, the UN requires effective means of communication between air and ground.

COMMUNICATION: MORE THAN JUST TALKING

Beyond regular communication between pilots and air traffic control towers, this facet of air power includes a wide array of other communication roles, broadly defined, with applications in peace operations. UN aircraft sometimes need to communicate, with voice or data, to ground military forces, whether to pass on their aerial observations and imagery or to coordinate close air support. But communications systems used by the troop-contributing countries or host-nation forces often are not interoperable with systems used by the aircraft. In the DRC, for instance, Indian Mi-35 (NATO term "Hind-E") helicopters could not communicate with the DRC forces during close air support, so the troops had to be previously instructed to build fires and use white flags to allow the helicopters to distinguish their positions from rebel ones when the helicopter fired missiles.[29] Though primitive and limited, the visible method worked in these instances, though it did not allow for sophisticated maneuvers.

When a particular nation supplies both air and ground military units to the UN or the air and ground forces have common standards, as do most NATO countries, it is easier to build on the strong ground-to-air links. Using these strong links between its aircraft and troops, South African commanders in the UN mission in the DRC would remain heliborne to view battles and guide the troops. Thus, command and control (C2) is combined with communications to be more effective C3.[30] In other cases, the UN purchased dedicated equipment (for example, Jotron VHF/UHF radios) for air-to-ground communication with contracted commercial aircraft.

Communication with the ground is not always done by voice or data. For a show of force, an aircraft can "buzz" a location by flying low overhead, whether to intimidate an adversary or to reassure friendly forces. The value of the airborne "presence" can be quite strong for UN ground forces in that they now know that aircraft are supporting them from above. Lieutenant-General (ret.) Roméo Dallaire, force commander of the UN Assistance Mission for

Rwanda during the genocide, noted that the sound of incoming UN aircraft provided immense reassurance and motivation well before the life-saving supplies arrived.[31] German soldiers on patrol in Mali would ask for Heron UAVs to circle above to alert them to any threats and help coordinate ground movements. Seeing the UAVs in the sky gave them a sense that if they encountered difficulties, the UAV could provide situational awareness to the sector headquarters as well as to themselves because the UAV image feed was viewable on their handheld terminals (which are generically called Remotely Operated Video Enhanced Receivers, or ROVERs).

Aircraft, including UAVs, can also communicate to the ground by light or laser. Searchlights from helicopters can illuminate areas for better observation and detection and can let persons on the ground know they are being observed or followed. Laser Range Finders (LRFs) can measure the distance from the aircraft to a person or object and hence determine their position. Laser designators can pinpoint targets for ground forces and aerial precision-guided munitions.

Lasers can be used not only for observation and targeting but also for communication. Though not used in peacekeeping yet, this option is worth exploring. LRFs usually operate in the invisible part of the electromagnetic spectrum, but some LRFs can be tuned so their beams are visible to the human eye (while still being eye safe), which means they can be used to signal a person that he or she is being watched or even targeted at night. For instance, a laser beam can be focused on the person's chest or in his or her immediate vicinity. A perpetrator engaged in illegal acts or sexual attacks might be convinced to cease the illegal activity when so targeted by an intense red dot or other marker. Alternatively, rebel forces moving to attack can be warned by laser signals aimed at them or by "writing" on the ground that communicates spatial limits—"red lines"—or even by symbols to instruct them to stop their advance. This function has not yet been developed in peace operations, but with the advancement of technology it may play a prominent role in that arena in the future.

Simpler technologies have already been used for communications. Instead of light and lasers, sound can convey messages from aircraft, in particular helicopters. Heliborne loudspeakers have been used to hail persons on the ground. The voice can provide reassurance, give instructions, or warn/threaten wrongdoers. When humanitarian aid is being landed by helicopter, aircrews have ordered locals to move away from the landing area, especially if no ground forces are in place to secure the landing zone. The Israeli Quadcopter Hover Mast, which is tethered to a slow-moving pickup-truck, was deployed to the UN mission in the Central African Republic. In addition to observation from 100 meters it also makes a very loud noise, which UN force commandars used as a show of force to locals. The UN might even explore the options for airborne nonlethal weapons, including sound-producing devices designed to cause irritation and weapons that fire blunt-impact projectiles or contain indelible paint. Road spikes (caltrops) can be dropped from the air to temporarily halt a vehicle or convoy moving to attack a village.

Leaflets have also been dropped from aircraft to help the UN communicate with local persons, including hard-to-reach combatants hiding in remote or forested areas. For instance, MONUC helicopters flew over the strongholds of rebel groups to drop leaflets to encourage them to participate in UN disarmament, demobilization, rehabilitation, reintegration, and repatriation programs.[32] Similarly, in Haiti, a UAV was used by a Brazilian battalion in 2006 to drop leaflets to encourage gang members in Cité Soleil to surrender before a forceful UN intervention and other leaflets to warn locals to stay indoors during impending operations.

Aircraft can also be used to relay signals to extend the range of communication between ground forces or to assist other aircraft. Because most radios (VHF and UHF) operate on a "line of sight" basis, they cannot communicate over distances longer than 3.1 to 6.2 miles (5 to 10 kilometers) due to the earth's curvature, and the range is also highly dependent on terrain. Natural features such as mountains can obstruct signals. Because of the large volcanos near Goma, DRC, the UN peace operation used one UAV to serve as a relay vehicle to communicate to a second UAV beyond the volcano. Otherwise, the signals from the ground control station would be unable to reach the second UAV. This relay almost doubled the range of flight and observation.

For areas with no internet connectivity because of a natural disaster, armed conflict, or simply underdevelopment, it is possible to provide internet access over an area through aircraft, particularly long-endurance UAVs or aerostats. This could be useful to civilian and military peacekeepers who need internet connectivity in remote locations, including when preparing for UN combat operations or engaging in them.

COMBAT: ATTACK HELICOPTERS FOR PEACE

The idea of "attack helicopters for peace" may seem oxymoronic or at least incongruous, but these advanced weapons systems have become an important part of several modern peace operations. Attack helicopters (AHs) have been used in this century to enforce peace and to protect peacekeepers and civilians in Sierra Leone, Liberia, Côte d'Ivoire, the DRC, Darfur, and Mali.

The Congo, which had seen UN combat aircraft, mostly jets, in the 1960s, was the scene of another UN aerial innovation in the twenty-first century: the incorporation of AHs. The UN's acquisition of an Indian Mi-35 helicopter unit in 2003 allowed for the UN mission, MONUC, to use force more robustly. These helicopters have provided much needed area reconnaissance and, in some cases, close air support with rockets and machine guns. However, due to safety reasons, these helicopters were not sent out at night until 2006. The Mi-35's Forward-Looking Infrared Radar and the crew's night-vision goggles allowed the helicopters to fly on night missions, which were permitted on the basis of urgent need. The flights also provided the actionable intelligence needed about hostile militia forces, who often moved at night and who had previously been confident that the UN aircraft would not be operating then.

Despite the successes, the Mi-35 deployment still had limitations. These helicopters were able to stay on station for only 1.5 hours maximum before they had to refuel. Extreme weather conditions and communication delays with ground forces sometimes resulted in lost opportunities because helicopter crews could not engage against militia before the latter slipped into the jungle.[33] Also, the UN limited helicopter night flying even after 2006 because it lacked a night search-and-rescue capability that would be needed in the event of a crash. Militia groups learned that they could usually operate unimpeded at night, leading to the situation of peacekeeping by day and rebel pillaging by night. The advent of UAVs removed the dangers of night flying for pilots, but UN UAVs only carry out surveillance and cannot land troops or fire against hostile forces.

In 2013, South Africa also deployed three Denel Rooivalk combat-support helicopters to complement the Mi-24 flown by Ukraine, which had taken over from the Indian unit of Mi-35 helicopters (enhanced export version of the Mi-24). On 4 November 2013, two Rooivalk helicopters engaged in their first combat mission against the notorious rebel group M23 near Chanzu, a mountainous region close to the Rwandan border.[34] The helicopters fired their 70-millimeter rockets at the bunkers of M23 and also took out one of M23's 14.5-millimeter antiaircraft guns that was probably used against other UN helicopters in previous engagements.

Attack helicopters proved to be a vital component in MONUSCO's operations. Combat capability and air superiority over rebels are only two aspects of this advantage, with the sensors observing violence, smuggling, and other illicit activities that perpetuate conflict in the DRC. And with the successful use of AHs in the DRC, the UN subsequently deployed them to Darfur, Mali, and the Central African Republic. The latter offers a powerful case study of their use, including some of the tactical aspects.

ATTACK HELICOPTERS IN THE CENTRAL AFRICAN REPUBLIC

In 2013, the Séléka group, a predominantly Muslim coalition, overthrew the Central African Republic government of President François Bozizé. Faced with immense internal and external pressures, including a French military intervention, the new government collapsed. Within months, the Séléka forces disbanded, but many of its members did not lay down their weapons. By 2014, the country was partially divided between ex-Séléka members predominant in the northeastern part of the country and the Christian-majority anti-Balaka militia in the southwestern part. Ex-Séléka forces not only fought with anti-Balaka forces but also among themselves. Into this confusion, MINUSCA was established in 2014 to help bring peace and protect civilians.

On 21 January 2017, a UN patrol spotted 150 armed militiamen from the Front populaire pour la renaissance de Centrafrique (FPRC), one of the most powerful ex-Séléka armed groups in the Central African Republic, whose forces were moving to attack the town of Bambari. The UN mission instructed the FPRC fighters to stop their movement westward based on a previously declared UN "red line." Specifically, one part of the red line was declared on the Bria–Ippy–Bambari axis. The warning was successful in that case, and the fighters turned back.

One month later, however, contrary to UN demands to cease offensive actions, on 10–11 February the FPRC moved a substantial attack force of about 300 men in a column toward Bambari. Armed with automatic weapons (AK-47) and rocket-propelled grenades, they were assisted by seven pickup trucks in their movement along the Ippy–Bambari road. Their goal was to attack and sack Bambari and remove the leader (Ali Darassa) of another ex-Séléka group based there. Seeing that the UN's red line had indeed been crossed, MINUSCA commanders sent a Senegalese Mi-35 AH helicopter on

11 February to a location near the village of Ngawa, where it observed the rebels' movement using the Mi-35's high-zoom cameras in a pod at the front of the aircraft. The pod could observe in both the visible and the infrared parts of the electromagnetic spectrum, and the cameras were gyrostabilized to reduce vibration. The images were used for positive identification of armed/uniformed personnel on the ground and for more precise targeting.[35]

The MINUSCA force commander, General Balla Keïta of Senegal, had already signed a fragmentary order to allow for the Mi-35 to engage, with tactical control assumed by the mission's Joint Task Force in Bangui. After firing a warning shot, the Mi-35 fired on the armed convoy with rockets and machine-gun rounds, destroying four pickup trucks and scattering the rebel forces. The rebels probably fired at the helicopter with their semiautomatic rifles, but on the helicopter's return to base in Bangui, a postflight check found no bullet holes in it.[36] The UN action had successfully stopped the rebel movement to Bambari. Most of the rebels retreated to Ippy. There were an unknown number of fatalities, but four injured people were brought in to the Ippy health center. UN military observers received a report that the man leading the FPRC column, "General" Joseph Zoundeko, was killed during the incident.

Despite the blowback from the FPRC in response to this action, MINUSCA demonstrated that it could protect civilians threatened with imminent attack. In order to show impartiality and further protect the people of Bambari, General Keïta ordered a show of force to pressure Union for Peace in the Central African Republic (UPC) leader Ali Darassa to move out of the town. MINUSCA did not want any fighting in the town between the opposing rebel groups, especially fighting that might kill civilians. So MINUSCA surrounded the rebel leader's compound and stationed the Mi-35 directly overhead, forcing Darassa to leave.[37] Later, when Portuguese and UN drones detected the movement of thirty-forty UPC fighters attempting to attack the mission's Rapid Reaction Force (Portuguese), the force engaged the rebels and destroyed several trucks.[38] Thus, both conflicting parties learned that a UN "red line" was meaningful and backed by force, thanks in part to the presence of the Senegalese Mi-35 AHs and UN UAVs.

This case again showed the utility of AHs for peace enforcement within the context of a peace operation. Though studies of twentieth-century peace operations by Trevor Findlay concluded in 2002 that air combat power is of marginal use in peace operations,[39] the UN's practice and experience in the twenty-first century proves that finding wrong. Sophisticated AHs have been deployed to great effect in a number of UN missions. In the DRC, other nations deployed AHs, including South Africa with its Rooivalk AH. In Mali, the Netherlands deployed the Apache AH, and Germany deployed the Tiger AH. Canada later deployed its Griffon helicopters (with machine guns in the doors) to escort the medium-transport Chinook aircraft. Also, El Salvador contributed the light MD-500 helicopters equipped with a machine gun and missiles. Attack helicopters brought into peace operations by nations from Europe, Asia, and Africa have proven effective, though more case studies are needed on the use of and experiences with AHs and combat air power more generally.

CONCLUSIONS, LESSONS, AND RECOMMENDATIONS

Despite some successful instances of UN air power application, especially combat power, there are many instances where the United Nations has not used enough force or responded too late or not at all to atrocities. In cases such as Bosnia and Rwanda from 1993 to 1995, the UN lacked not only the combat capability but also the means to effectively communicate between air and ground forces. Similarly, the UN's airborne observation capability, though enhanced by a modest fleet of UAVs since 2013, has not been sufficient for early warning and for many robust actions. Also, air transport within vast and war-torn countries remains an operational challenge, despite safety improvements in the twenty-first century. Finally, the world organization has found it difficult to move from a reactive approach to a proactive one.

UN member states often provide only tenuous and short-term support with equipment (including military aircraft) and air personnel. Interoperability is difficult given the eclectic mix of commercial and military aircraft the UN gathers from the developed and developing world. Considering these disadvantages, it is surprising that the UN's operations have helped establish peace in dozens of countries, thanks in large part to the air military and civilian staff willing to accept personal risks and hardships in the field for the greater good.

There is much more to do to improve UN operations in each of the four facets of air power. For transport, the UN needs more and more modern aircraft. Although it has spent roughly $1 billion per year on aviation in its peace operations for the past decade, it still relies primarily on Russian-built or contracted aircraft, in particular the ubiquitous Mi-8 helicopters flown by Russian/Ukrainian contractors and developing-world militaries. Russia is careful to guard its contracts, and the UN system is biased toward Russian aircraft because winning the bidding process means providing air services that meet the requirements at the lowest price. To bring in more options for contracted aircraft, the aircraft should be scored for capabilities, including those not in the requirements list. The UN might even consider purchasing some of its aircraft rather than simply leasing them, given the organization's high demand for and steady usage of such aircraft. This would allow more rapid deployment and greater inhouse expertise and may be less expensive in the long run, given that the UN has a constant need, which contractors and nations sometimes have a hard time fulfilling. The UN has already gained ownership of many micro-UAVs. It could expand its ownership to larger UAVs and even manned aircraft.

New forms of transport can also be envisaged. For instance, parafoils guided by the Global Positioning System can be dropped from aircraft and fly autonomously or by remote pilot to reach their destination to deliver supplies and humanitarian aid. This has some benefits over manned delivery in areas with hostile actors, poor visibility, and night-flying prohibitions for manned aircraft. Airships are a potential future platform, but they need to be proven first in the general aviation industry.[40]

Another innovation is to provide aircraft that can serve multiple operations at the same time rather than being assigned to only one country or mission. This requires empowering a tasking authority, probably the Strategic Air Operations Center at the Regional Services Center in Brindisi, Italy. Canada is pioneering this possibility by providing a C-130 Hercules for five days a month to serve multiple UN missions out of Entebbe, Uganda, where the UN already has a major logistics hub. The aircraft is helping several UN operations on an urgent-need basis, from DRC to South Sudan.

In this age of information- and intelligence-led operations, the United Nations needs more observation aircraft as well as better cameras. But the greatest need is for reliable data access and deeper analysis. With the expanding flow of imagery from UAVs and other platforms, the world organization needs systems to store, access, and analyze that imagery, including the capacity for multispectral and synthetic radar aperture analysis as well as infrared signature analysis. To catch perpetrators, UN air patrols cannot remain limited to the daytime. Manned night flights and UAVs offer the possibility of coverage at times when nefarious actors are most likely to carry out smuggling and killings.

The UAS revolution offers the distinct advantages of longer loiter time, greater pilot safety, and real-time image transmission. In addition, smaller UASs allow the ability to use man-portable and man-deployable aircraft for tactical reconnaissance. Technology-contributing countries (abbreviated as TechCCs) need to provide UAS with qualified military and civilian personnel for piloting, payload operation, and effective analysis of the airborne imagery.

The United Nations needs interoperable communications systems, especially for air and ground units that are engaged in combat together. This is surprisingly not the case for the three countries in the Force Intervention Brigade (Malawi, South Africa, and Tanzania), which has a unique mandate for "offensive" operations against rebel groups in the eastern DRC. The UN can specify some basic standards for communications and use equipment, such as Radio Interoperability Systems that can connect diverse communication systems, including mobile phones and aircraft radios. Such systems can also be used to pass data, including tracking information and imagery.

To increase the value of UASs, it is also necessary to connect in real time the UAS operators with the forces on the ground. This contact can potentially be done through the UAV itself, using chat functions or voice transmission. Thus, mission analysts can alert the forces quickly if they see threats—for example, the planting or location of improvised explosive devices. New forms of communication can also be explored, such as using UAVs or other aircraft to shine eye-safe laser beams to signal presence and thus let potential belligerents and criminals know that their actions are being watched and recorded. With the powerful lasers now available, it should be possible also to "paint" symbols on the ground for more sophisticated messaging.

Whereas the first fifty years of peacekeeping saw combat aircraft firing only in the Congo, the twenty-first century has witnessed the rise of the attack helicopter as the symbol of robust peace operations. These remarkable machines have proven their worth on numerous occasions, but detailed studies and lessons of their use are still needed. For the UN to respond rapidly to imminent threats to civilians, it is essential that AHs, or at least armed utility helicopters, become a feature of UN missions rather than the exception. The Western governments that have provided AHs in the past (Germany and the Netherlands) should be encouraged to provide descriptions of both lessons learned from their use and their capabilities for the benefit of other missions. Of course, UN missions will need to develop robust command-and-control structures to make effective use of these aircraft.

The use of jet aircraft in peace operations is more debatable because of the risk of collateral damage on the ground and the high expense for a world organization facing budgetary challenges. The speed and kinetic action that fighter and bomber aircraft bring to bear would likely strain any UN peace operation. Decision making within UN operations is typically too slow to make good use of fast jets. And the offensive capabilities brought to bear by means of a jet are often well beyond that typically used in peace operations, where small arms and light weapons are the standard weapon systems of conflicting parties.

Despite some impressive instances of UN air power in many more cases the UN did not use enough force or responded too late or not all during atrocities, as in eastern Congo (2004, 2012), South Sudan (2016), and Mali (2019).[41] For peace operations to be quick and proactive, improved capabilities and political will are needed.

The UN's procurement processes currently hold the organization back from rapid responses to emerging trends. Commercial procurement processes are quite slow, often taking more than a year and a half from the invitation to bid to the first aircraft in service. For emergencies, the UN peace operations need to do better. The World Food Programme and its Humanitarian Air Service have proven capable of rapid response. At UN Headquarters, procurement processes need improvement—for example, by taking bidders' past performance into account because some of them have poor records of fulfilling contracts. The bidding process itself should also look at value to be provided, not just at the lowest bid that meets the requirements. The UAS bidding process showed the way by allowing the bids to be evaluated on the basis of quality, including operational benefits beyond minimum requirements, as well as on the basis of cost. In short, the world organization needs more capability, aircraft, and expertise in each of the four facets of air power.

The United Nations has made significant progress in air power in its twenty-first-century operations. The world's future will give rise to many new conflicts and thus require new UN operations, especially as the earth-changing effects of climate change exact their toll. There will be a continuing need for UN air capabilities to provide transport, observation, communication, and combat power. Peace and humanitarian operations will be in great demand, and we can be thankful that the United Nations will be there to help when it can, assisted by air power.

NOTES

1. The author thanks Stewart Webb of DefenceReport.com for his excellent assistance with researching, summarizing, and drafting this chapter. Much gratitude goes also to those who reviewed drafts of the chapter: Ryan Cross, Cono Giardullo, Richard Goette, William March, Robert Owen, Danielle Stodilka, and, of course, the volume editor, Steven Paget.

2. A. Walter Dorn, preface to *Air Power in UN Operations: Wings for Peace,* ed. A. Walter Dorn (London: Ashgate, 2014), xxv (this entire volume is available gratis at http://unairpower.net). See also Alexandra Novosseloff, *Keeping Peace from Above: Air Assets in UN Peace Operations* (New York: International Peace Institute, October 2017).

3. The peacekeeping mission that was deployed in Korea from 1948 to the outbreak of the Korean War did not possess any aircraft, which contributed to its failure to see the warning signs of an impending attack on 25 June 1950.

4. For a more detailed review of the Congo case and for additional specific references, see A. Walter Dorn, "Peacekeepers in Combat: UN Fighter Jets and Bombers in the Congo, 1961–1963," in *Air Power in UN Operations,* ed. Dorn, 17–39.

5. I. J. Rikhye, *Military Adviser to the Secretary-General: UN Peacekeeping and the Congo Crisis* (New York: St. Martin's Press, 1993), 193.

6. Kevin Spooner, "A Fine Line: Use of Force, the Cold War, and Canada's Air Support for the UN Organization in the Congo," in *Air Power in UN Operations,* ed. Dorn, 42–43.

7. Dorn, "Peacekeepers in Combat," 22, 25.

8. R. C. Johnson, "Heart of Darkness: The Tragedy of the Congo, 1960–67," *Chandelle: A Journal of Aviation History* 2, no. 3 (1997): 11, also at http://worldatwar.net/chandelle/v2/v2n3/congo.html; Wayne Fredericks quoted in H. Klevberg, "Logistical and Combat Air Power in ONUC," in *Use of Air Power in Peace Operations,* ed. Carsten F. Ronnfeldt and Per Erik Solli (Oslo: Norwegian Institute of International Affairs, 1997), 36–37.

9. Rikhye, *Military Adviser,* 294–95; Trevor C. Findlay, *The Blue Helmets' First War? Use of Force by the UN in the Congo, 1960–64* (Clementsport: Canadian Peacekeeping Press, 1999), 118.

10. For details, see Dorn, "Peacekeepers in Combat," 26. Ethiopia abruptly withdrew its Sabre jets after a crash in 1962, and India withdrew the Canberra bombers due to the outbreak of war with China on 20 October 1962. In January 1963, the UN received additional Sabre jets from Italy, the Philippines, and Iran, although these jets arrived after Operation Grand Slam and did not engage in fighting before the end of the ONUC mission in June 1964.

11. General Kebbede to Dr. Ralph Bunche, outgoing code cable, 24 November 1962, S0829-1-14 (DAG 13/1.6.5.8.4.0), UN Archives, New York.

12. A. Walter Dorn and David J. H. Bell, "Intelligence in Peacekeeping: The UN Operation in the Congo 1960–64," *International Peacekeeping* 2, no. 1 (Spring 1995): 20–21.

More details about ONUC and references can also be found in Dorn, "Peacekeepers in Combat."

13. Quoted in Dorn, "Peacekeepers in Combat," 31.

14. Robert Owen, "Humanitarian Relief in Haiti, 2010: Honing the Partnership between the US Air Force and the UN," in *Air Power in UN Operations*, ed. Dorn, 77, 92.

15. A. Walter Dorn and Ryan W. Cross, "Flying Humanitarians: The UN Humanitarian Air Service," in *Air Power in UN Operations*, ed. Dorn, 102, 106.

16. Owen, "Humanitarian Relief in Haiti," 92, 95.

17. A. Walter Dorn, *Keeping Watch: Monitoring, Technology, and Innovation in UN Peace Operations* (Tokyo: United Nations University Press, 2011), 142–44.

18. Kevin Shelton-Smith, "Advances in Aviation for UN Peacekeeping: A View from UN Headquarters," in *Air Power in UN Operations*, ed. Dorn, 293–94.

19. The abbreviation MONUSCO stands for Mission de l'organisation des Nations unies pour la stabilisation en République démocratique du Congo, which was the new name for Mission de l'organisation des Nations unies en République démocratique du Congo (MONUC).

20. Shelton-Smith, "Advances in Aviation for UN Peacekeeping," XXX.

21. Colum Lynch, "U.N. Wants to Use Drones for Peacekeeping Missions," *Washington Post*, 8 January 2013, at https://www.washingtonpost.com/world/national-security/un-seeks-drones-for-peacekeeping-missions/2013/01/08/39575660-599e-11e2-88d0-c4cf65c3ad15_story.html.

22. Edmund Kagire, "Rwanda President: No Issue with UN Drones in Congo," *Washington Examiner*, 21 January 2013, at https://www.washingtonexaminer.com/rwanda-president-no-issue-with-un-drones-in-congo.

23. John Karlsrud and Frederik Rosén, "Lifting the Fog of War? Opportunities and Challenges of Drones in UN Peace Operations," in *The Good Drone*, ed. Kristin Bergtora Sandvik and Maria Gabrielsen Jumbert (New York: Routledge, 2017), 51.

24. "Joint INGO Position on Humanitarian Use of UAVs," *ReliefWeb*, July 17, 2014, at https://reliefweb.int/report/democratic-republic-congo/joint-ingo-position-humanitarian-use-uavs.

25. Ibid.

26. Other Heron-1 systems were already deployed by French forces in Mali; however, their use was restricted to the French counterterrorism mission Operation Barkhane, comprising about 3,000 French personnel. The Heron-1 has a maximum mission endurance of more than twenty-four hours, which is a significant increase on what a helicopter can endure.

27. Erwan de Cherisey, "Desert Watchers: MINUSMA's Intelligence Capabilities," *Jane's Defence Weekly* 54, no. 23 (7 June 2017): 6.

28. David D. Kirkpatrick, "Russian Snipers, Missiles, and Warplanes Try to Tilt Libyan War," *New York Times*, November 5, 2019, at https://www.nytimes.com/2019/11/05/world/middleeast/russia-libya-mercenaries.html.

29. Dorn, "Combat Air Power in the Congo," in *Air Power in UN Operations*, ed. Dorn, 241–53.

30. Personal communications from UN and South African commanders to the author during visits to the MONUSCO mission, July 2017.

31. Romeo A. Dallaire, foreword to *Air Power in UN Operations,* ed. Dorn, xxi.

32. A photo of such leaflet distribution by helicopter can be found in the UN photo library: photograph no. 202011, at http://downloads.unmultimedia.org/photo/ltd/high /202/202011.jpg.

33. Dorn, "Combat Air Power in the Congo," 253.

34. Darren Olivier, "Rooivalk Attack Helicopters Perform Well in First Combat Action against M23," *African Defence Review,* November, 5, 2013, at https://www .africandefence.net/after-23-years-rooivalks-fire-first-shots-in-drc/.

35. The Senegalese Mi-35 system has a Controp DSP-1 camera pod, which contains a high-definition visible-light camera, an infrared camera, and a Laser Range Finder to determine the distance to targeted objects.

36. Information on this case study is from UN staff involved in the use of force, especially the author's conversations with the Senegalese Aviation detachment commander on 27 March 2018, subsequent email exchanges, and documentation from the MINUSCA Joint Operations Centre, in particular "Daily Situation Report Covering Period: 9 February 2017 (0001–2400)," 10 February 2017, Sage (MINUSCA's incident and event database).

37. Author's conversation with General Balla Keïta, MINUSCA force commander, Montreal, 11 June 2019.

38. Fergus Kelly, "Portugal Paratroopers Counter Armed Groups around Bambari, Central African Republic," *Defence Post,* 21 January 2019, at https://thedefensepost .com/2019/01/21/central-african-republic-portugal-paratroopers-counter-armed-groups -bambari/.

39. Trevor Findlay, *The Use of Force in UN Peace Operations* (Oxford: Oxford University Press, 2002), 85, 268–69, 282.

40. Walter Dorn, Nic Baird, and Robert Owen, "Airships in U.N. Humanitarian and Peace Operations: Ready for Service?," *Journal of Aviation/Aerospace Education & Research* 27, no. 2 (May 2018): 41–52.

41. These cases are tragic failures in which UN forces were unable or unwilling to protect the local population during attacks—for example, the rebel attacks and temporary occupations of the Congolese towns of Bukavu (2004), Kiwanja (2008), and Goma (2012); South Sudanese government forces' massacre of civilians of certain ethnicities in July 2016; and in central Mali, killings sprees by both government and rebel forces in 2019. This chapter highlights some of the many successful cases where UN forces did protect civilians using air power.

Multinational Air Power

The Outlook

Steven Paget

Anthony Schinella observed: "For as long as men have taken to the skies, theorists and strategists have dreamed of using airpower to achieve victories or compel adversaries without having to close with the enemy and destroy him in a bloody, close-quarters fight."[1] For as long as humans have employed air power, furthermore, they have frequently done so in a multinational context. The Royal Air Force (RAF), the world's oldest independent air force, has a history of multinational operations that spans from its creation in the First World War to contemporary air policing operations in the Baltic. Air power can be, is, and will continue to be applied independently, but it will also be utilized frequently in a coalition environment. The RAF, for example, has been explicit about the need to "generate and sustain delivery of battle-winning air and space power for UK and Coalition Commanders."[2] Even when air power is applied independently, it may well be used against multinational adversaries, as in the Israeli experience during the Six-Day War.[3] Joseph Soeters and Philippe Manigart have contended pertinently: "Multinational military cooperation is here to stay. Like firms in international business, national armed forces are increasingly compelled to work together because of the frequency, nature and scale of the missions they are tasked to do."[4] As "coalitions of widely disparate cultures and countries are very much the order of the day," the benefits and challenges of multinational operations must be understood.[5]

Multinational operations provide a range of benefits. The formation of a coalition and operations conducted by multinational fora enable partners to "signal the credibility of their mutual commitment."[6] Political credibility is of great significance, but the potential benefits are much broader. Coalitions offer states a means both of "bolstering their capabilities" and of "spreading the cost of fighting among their allies."[7] Forming coalitions as a means of enhancing military capability has underpinned the provision of multinational air power. The US Army Air Forces' contribution to the Western Desert Air Force during the Second World War, for example, was essential for the RAF, not least because of the provision of American-built aircraft. From a practical and financial perspective, aside from the obvious savings through the deployment of other nations' militaries, access to forward-operating bases in a partner's country, as demonstrated during the Gulf War, among other conflicts, is extremely valuable.[8] The benefits inevitably also come with costs.

Multinational coalitions—even when successful—have faced a range of challenges that result from the need to integrate. Michelle Pryor, Thomas Labouche, Mario Wilke, and Charles Pattillo Jr. have warned cogently: "Whether the operation involves an established alliance or an ad hoc coalition, interoperability between multinational forces is imperative to achieving mission success. To be successful in the anticipated complex and shifting operating environment, coalition forces must identify and address potential strategic and operational challenges and interoperability concerns well in advance."[9]

That maxim, although both logical and correct, has proven to be easier said than put into practice throughout the history of multinational operations. It has been noted: "Achieving and sustaining interoperability among diverse national military systems is not easily attainable and depends on technological, logistical, and human factors."[10] It has also proven to be a truism, moreover, that "determining whether a given set of countries are [sic] interoperable or not is problematic . . . as it is difficult to establish exact standards for when the multinational cooperation should be considered effective."[11]

The challenges that arise during multinational operations present themselves at the strategic, operational, and tactical levels. The levels of war—understood as a hierarchical construct generally—are, in practice, more like an overlapping Venn diagram. John Kiszely has written that "few decisions facing senior military commanders in multinational campaigns do not have a significantly political dimension, with implications for political involvement and control."[12] When it comes to rules of engagement (ROE), for example, decisions at the strategic level will be implemented at the operational level, where a red-card holder will be positioned to prevent any departure from the imposed limitations. The front line, however, effectively becomes the last line of ensuring that ROE are upheld because it is the aircrew who will ultimately need to prevent infringements. The strategic corporal concept, in which decisions taken at the tactical level can have consequences at the strategic level, is just as, if not more, applicable to aircrew.[13] The complexities of multinational air operations are interconnected and cannot easily be isolated. The different aspects of interoperability can be viewed as a series of interlocking gears that interact together to drive multinational operations. Those metaphorical gears require lubrication and adjustment to ensure the smooth running of multinational operations.

The ability of different nations' air forces to integrate is key, and limited jointness among national services can detract from the conduct of multinational operations. Friction between the French Army and Air Force during the Battle of France was not the cause of the limited interoperability between the RAF and the Armée de l'air, but unresolved tension over how the French would employ air power did make cooperation more difficult. Multinational partners can, however, sometimes aid counterparts when jointness is lacking. During the Iraq War (or Second Gulf War) of 2003, for example, US Navy and Marine Corps aircraft were not always compatible with US Air Force (USAF) tankers that used the boom system for refueling, but they could replenish from RAF tankers. Even when high levels of jointness exist within national militaries, the intricacies of multinational cooperation can present obstacles that need to be overcome.

Multinational operations have been conducted by air forces employed in a variety of structures, from "coalitions of the willing" to multilateral organizations. The North Atlantic Treaty Organization (NATO), which has been regarded as a "predominantly military organization," has taken great steps to improve interoperability, but it has not eradicated all points of friction.[14] Indeed, NATO has been labeled an "alliance à la carte divided into two or more factions of member states with divergent interests."[15] This division has had an inevitable effect on the formulation of strategy, but it has also been reflected in the differing levels of interoperability achieved by NATO's members. Operation Unified Protector in Libya in 2011, for example, revealed "serious deficits among European allies in conducting modern military campaigns."[16] The diverse membership of the United Nations and the complexities of its operations have ensured that integration has been a challenge in the conduct of air operations under the organization's auspices. Therefore, although the involvement of multinational organizations in an operation offers a number of benefits and their efforts have resulted in positive progress, it should not be assumed that interoperability will not be a challenge.

"Coalitions of the willing" have become more common, and as James Cook has reasoned, such coalitions come with advantages that formal organizations may not benefit from: "Less formal security arrangements offer the advantage of greater flexibility and largely avoid the criticisms of a 'war by committee' approach that was on display during the 1999 NATO Operation Allied Force."[17] Although an informal relationship can provide more flexibility, both the potential absence of common doctrine and the presence of limited interoperability can hinder multinational operations. The ability to integrate and, subsequently, effectiveness and efficiency may come down to how well the cooperating air forces know each other, irrespective of the structure in which they are operating.

Coalitions are not uniform, and the challenges to integration can be varied, most notably in the realms of cultural, human, procedural, and technical interoperability. In general, the challenges are likely to increase in both scope and depth as the diversity of the coalition increases. The air forces of regular multinational partners, such as the "Five Eyes" (Australia, Canada, New Zealand, the United Kingdom, and the United States), share a common ethos and language, which does not negate problems but does help to simplify integration. Language, for example, can be a challenging point of friction, as demonstrated to differing extents by the operations of the RAF and Armée de l'air during the Battle of France in 1940 and operations involving Vietnamese Air Force forward air controllers and Royal Australian Air Force (RAAF) 2 Squadron Canberras in Vietnam. Indeed, human and cultural factors have proven to be of great significance during the conduct of multinational operations.

The value of personal relationships, for example, should not be underestimated. The close professional relationship maintained by Lieutenant General T. Michael Moseley, USAF; Air Vice Marshal Glenn Torpy, RAF; and Group Captain Geoff Brown, RAAF, during Operation Iraqi Freedom was reminiscent of the setup experienced by Wing Commander David Evans when, as 2 Squadron (RAAF) commanding officer during the Vietnam War, he was treated as the senior national Australian officer by Colonel Frank

Gailer Jr., USAF, who served as commanding officer of the 35th Tactical Fighter Wing. Multinational-minded officers who are proactive, forward leading, and, ultimately, willing to "evaluate how a member of a cultural group . . . sees himself, determine what is important to him, and consider it when developing operational objectives, goals, and methods to achieve these operational goals" are essential for the conduct of multinational operations.[18] The forward-leading approach adopted by Lieutenant General Charles Horner, USAF, to international partners during the Gulf War exemplified the proactive approach required for multinational operations.

Procedural interoperability, which "focuses on doctrine and procedures from the strategic, national level to tactical-level execution,"[19] has the potential to ease integration and reduce friction during the conduct of operations. The use of translated French doctrine and adoption of French organizational and staff structures by the American Expeditionary Forces Air Service during the First World War provide an early example of the benefits of procedural interoperability. The work of multinational fora such as NATO to standardize tactics, techniques, and procedures is evidence of the ongoing importance of efforts to enhance procedural interoperability. As multinational air forces become more familiar to each other, integration—although never without complications—does become easier, as demonstrated by post–Cold War operations involving the RAF and USAF.

Technical interoperability can be defined as "the mechanics of system technical capabilities and interfaces between organizations and systems. It focuses on communications and computers but also involves the technical capabilities of systems and the resulting mission compatibility or incompatibility between the systems (hardware and software) and data of coalition partners."[20] Achieving technical interoperability can be extremely expensive, but it can also raise issues of national sensitivities due to the potential to reveal technological secrets or classified information. Hesitancy over the provision of sensitive technology relating to unmanned aerial vehicles during UN operations has been a notable example of the barriers to integration in the technical realm. The differing reactions to the Defense Capabilities Initiative after Operation Allied Force in Kosovo, which some less-affluent nations considered exclusionary, emphasized that economic factors can be extremely limiting in achieving technical interoperability.

National caveats (the limitations a nation imposes on its armed forces) and ROE, which have become increasingly important throughout the history of multinational air operations, are determined by individual nations' perspectives. It has been posited that "among coalition Force Commanders, caveats are seen mainly as a severe impediment to military flexibility and efficiency, and thus to the successful implementation of the political mandate of coalition forces."[21] Per Marius Frost-Nielsen has gone even further, arguing that "caveats hamper coalition commanders' operational flexibility and often require coalition forces to fight with one hand tied to their back—potentially reducing the coalition forces' military progress."[22] Throughout the history of multinational military cooperation, notable differences between partners have had the potential to impede the effective conduct of operations. Target selection, for example, proved to be a controversial issue during Operation Allied Force, and the involvement of some air forces in actual combat operations was delayed during Operation Desert Storm due to national considerations. Remaining pragmatic during the course of operations, demonstrating understanding attitudes, and tasking forces carefully have proven to be essential to preserving coalition unity, even if effectiveness has been compromised to a degree.

Logistics was a vital issue during multinational operations above the trenches in the First World War, in the Western Desert in the Second World War, over the jungles of South Vietnam, and throughout various post–Cold War conflicts in the Middle East. Logistics has been labeled a "pivotal component of war which, on the one hand, is the arbiter of strategic opportunity and, on the other hand, is heavily determined by strategic and operational planning."[23]

Although the issue of dependence sometimes dominates the debate, logistics in multinational operations is not solely the concern of the preeminent contributor to an operation. It has been noted, for example, that the Italian contribution to logistic support of operations in Libya in 2011 was extremely significant: "Crucially, Italy made seven ITAF [Italian Air Force] bases available to NATO for Operation Unified Protector and—less visibly, but perhaps more importantly—provided the comprehensive connecting tissue of infrastructure, logistics, consumables, and services that Italy was asked to provide as host nation."[24]

Multinational operations are a team effort, as typified by the fact that more than 90 percent of fuel used in Operations Desert Shield and Desert Storm were reportedly provided by non-US members of the coalition. Of equal importance, although some nations such as the United States have

Two USAF F-16 Fighting Falcons refuel from an RAAF KC-30A Multi-Role Tanker Transport aircraft during Exercise Pitch Black in 2018. USAF photograph by Senior Airman Savannah L. Waters, at https://www.af.mil/News/Article-Display/Article/1607790 /air-force-improves-interoperability-strengthens-indo-pacific-partnerships/.

been notoriously generous, it is not solely benevolence that drives logistical support to coalition members. Providing such support is also an opportunity to enhance the effectiveness and safety of coalition operations, which is beneficial for all air forces involved in the conduct of multinational endeavors. Logistics will, nevertheless, as Roger Palin has argued, "remain essentially a national responsibility," and, as a result, it is important to "determine the degree of multinationality achievable."[25] Geography is sometimes key. Operations in Iraq and Libya, for example, have ensured that the support of local partners has been essential. Conversely, in Vietnam, 2 Squadron, RAAF, required significant US logistic support, but the existence of a financial working arrangement prevented any tension. Logistics has been a complex issue throughout the history of multinational operations, but its centrality to the success of operations is abundantly clear.

The challenges are myriad, but they can be overcome or at the very least mitigated to ensure that the benefits of coalition operations can be reaped.

USAF F-35A Lightning IIs (*middle*) flanked by RAF F-35B (*left*) and Israeli Air Force F-35I (*right*) work on multinational interoperability during Exercise Tri-Lightning in June 2019. USAF photograph by Staff Sergeant Keifer Bowes, at https://www.af.mil /News/Article-Display/Article/1887562/three-nation-f-35-exercise-demonstrates -air-power-interoperability/.

Practice might not always make perfect, but multinational exercises have been indispensable in increasing interoperability between regular multinational partners, as demonstrated by Operation Desert Storm. Major and regular exercises such as Red Flag dominate the headlines, but all exposure to partner air forces helps to improve the capacity to cooperate.

As Maria Burczynska points out, "The more frequently different air forces work together the better the opportunity to gain further understanding of each other's values, experiences and practices."[26] The benefits of exercises are manifold and traverse the different aspects of interoperability. Differences in doctrine as well as in tactics, techniques, and procedures can be resolved, and technical incompatibilities can be rectified. The human and cultural interoperability elements to exercises also help to break down barriers and build bridges. Participants are likely to emerge with a better understanding of the attitude, culture, and values of the members of their counterpart air forces. In some instances, particularly in the case of more

senior officers, lasting personal relationships can be developed that will ultimately pay dividends during the conduct of operations.

Preemptive measures are clearly preferable to reactive action, but the
conduct of operations exposes breakpoints that need to be addressed in theater. The work of the USAF's "FAC University" (504th Theater Indoctrination
School) in Phan Rang during the Vietnam War exemplified the need for wartime exigencies to help smooth the integration of multinational air forces.
The precombat training conducted by multinational air forces prior to Desert
Storm further emphasized the necessity of familiarization. Sustained interaction in the region in the period between Desert Storm and Operation Iraqi
Freedom, which was punctuated by Operation Desert Fox and Operations
Northern Watch and Southern Watch, provided a bedrock for cooperation
between the RAF and the USAF. Although the scale is obviously much
smaller, exchange and liaison officers offer another means through which
interoperability can be "developed and enhanced" during and before the conduct of operations.[27]

Multinational cooperation and collaboration have provided air forces, in
particular smaller ones, with opportunities to increase their operational
capacity. The growth of the American Expeditionary Forces Air Service with
support from France, Italy, and the United Kingdom during the First World
War and the provision of US aircraft to RAF Middle East during the Second
World War are prominent examples. Cooperation between unequal partners
has, however, reduced flexibility at times. In the case of the Hungarian Air
Force during the Second World War, assistance from the Luftwaffe fostered
dependence and left the Hungarian service an auxiliary of its German counterpart. Although a less-pronounced instance, the reequipping of the Royal
New Zealand Air Force with US aircraft enabled it to make a greater contribution to the war against Japan, but the provision of naval-type aircraft led to
its marginalization at times in the latter stages of the Second Word War. Support from other nations has been essential for some air forces in the effective
conduct of operations, but when assistance deteriorates into dependence,
then autonomy comes into question.

In assessing the worth of a given multinational partner, the emphasis is
on that partner's ability to add value to a particular operation. That addition
of value may be through an increase in mass or the inclusion of niche capabilities, such as the RAF's JP233 during Operation Desert Storm or the European Union Force's deployment of unmanned aerial vehicles to support the

UN operation in the Democratic Republic of the Congo in 2006. There is also a focus on limiting costs. Higher levels of self-sufficiency or, at least, a nation's willingness to pay its own way helps to foster goodwill. The size of an air force's contribution to a multinational operation is not the sole arbiter of its worth. The value of quality rather than simply size has been emphasized by the phrase "it's the chemistry not the physics" when British officers describe the United Kingdom's defense relationship with the United States.[28]

Technical interoperability helps air forces to plug into coalitions, and procedural interoperability enables the smooth conduct of operations. Strong human and cultural interoperability, developed through shared heritage, experiences, and training, aid cohesion and help to diffuse differences that arise during the conduct of operations. A multinational mindset is essential for coalition commanders to ensure that they understand and mitigate any issues that result from the integration of different national air forces. As part of that mindset, a process of mutual accommodation is required. The junior partners in any coalition should be cognizant of the overarching objective and seek to add value wherever possible. The senior partner must equally seek to understand the perspectives of its junior counterparts and be prepared to accommodate the limitations of other contributing air forces and maximize the benefits they bring to the endeavor. A competitive spirit, as characterized by the Royal Navy and Royal Australian Navy aircraft carriers' desire to increase their total number of sorties continually during the Korean War, can aid in a coalition's effectiveness if properly harnessed and kept under control.

Multinational operations have been a frequent feature of the use of air power since the birth of flight, and they are likely to become more prevalent. Joseph Soeters and Irina Goldenberg have contended: "National manpower and material resources are generally insufficient to address the demands of missions worldwide. The desire to optimize the use of scarce research and development and investment capabilities, the need for international legitimacy and political support, and the plain fact that today's risks do not end at national borders, have rendered multinational cooperation in the security domain unavoidable."[29]

Interoperability, much like military readiness, is perishable and must be enriched continually. Even then, the trajectory is not always forward. As barriers to cohesion are resolved, new problems—resulting from a range of challenges, such as the introduction of new technology and the need for cooperation with less-familiar partners—emerge and require resolution. Improvements in interoperability must be sought continually to enhance the efficiency and effectiveness of multinational cooperation. There is no silver bullet to the challenges presented by multinational operations, and obstacles of different natures must be both expected and resolved. Multinational operations may significantly boil down to what in the context of Australia's defense relationship with the United States has been termed "trustworthy mateship and reliable partnership."[30] Limitations will always be imposed on any multinational partner—whether in the form of more restrictive ROE or of technological shortcomings—but what is valued is the willingness to contribute as much as possible to the multinational mission within national constraints.

Notes

1. Anthony M. Schinella, *Bombs without Boots: The Limits of Airpower* (Washington, DC: Brookings Institution Press, 2019), 1.

2. Royal Air Force, *Royal Air Force Strategy: Delivering a World-Class Air Force* (High Wycombe, UK: Air Media Centre, 2017), 22.

3. The conflict is also referred to as the Arab–Israeli War of 1967 and the Third Arab–Israeli War. For more on the employment of air power in this conflict, see Alan Stephens, "Modeling Airpower: The Arab–Israeli Wars of the Twentieth Century," in *Airpower Applied: US, NATO, and Israeli Combat Experience*, ed. John Olsen (Annapolis, MD: Naval Institute Press, 2017), 217–84, and Noam Hartoch, "A History of the Syrian Air Force, 1947–1967," PhD diss., King's College London, 2015.

4. Joseph Soeters and Philippe Manigart, epilogue to *Military Cooperation in Multinational Peace Operations: Managing Cultural Diversity and Crisis Response*, ed. Joseph Soeters and Philippe Manigart (Abingdon, UK: Routledge, 2008), 223.

5. Peter Caddick-Adams, *Monte Cassino: Ten Armies in Hell* (Oxford: Oxford University Press, 2013), 373.

6. Marina E. Henke, *Constructing Allied Cooperation: Diplomacy, Payments, and Power in Multilateral Military Coalitions* (Ithaca, NY: Cornell University Press, 2019), 4.

7. Benjamin A. Valentino, Paul K. Huth, and Sarah E. Croco, "Bear Any Burden? How Democracies Minimize the Costs of War," *Journal of Politics* 72, no. 2 (April 2010): 32.

8. Sarah E. Kreps, *Coalitions of Convenience: United States Military Interventions after the Cold War* (Oxford: Oxford University Press, 2011), 40.

9. Michelle L. Pryor, Thomas Labouche, Mario Wilke, and Charles C. Pattillo Jr., "The Multinational Interoperability Council Enhancing Coalition Operations," *Joint Force Quarterly* 82 (3rd Quarter 2016): 112.

10. Jarle Eid, Francois Lescreve, and Gerry Larsson, "An International Perspective on Military Psychology," in *The Oxford Handbook of Military Psychology*, ed. Janice H. Laurence and Michael D. Matthews (New York: Oxford University Press, 2012), 114.

11. Joakim Erma Møller, "Trilateral Defence Cooperation in the North: An Assessment of Interoperability between Norway, Sweden, and Finland," *Defence Studies* 19, no. 3 (2019): 236.

12. John Kiszely, "The Political-Military Dynamic in the Conduct of Strategy," *Journal of Strategic Studies* 42, no. 2 (2019): 246.

13. Kersti Larsdotter has outlined that "what seems like tactical level decisions for the military might be considered major strategic decisions for the politicians, increasing the friction between the two" ("Military Strategy in the 21st Century," *Journal of Strategic Studies* 42, no. 2 [2019]: 162).

14. Sean Kay, "What Went Wrong with NATO?," *International Affairs* 18, no. 1 (2005): 77.

15. Timo Noetzel and Benjamin Schreer, "Does a Multi-tier NATO Matter? The Atlantic Alliance and the Process of Strategic Change," *International Affairs* 85, no. 2 (2009): 211.

16. Ellen Hallams and Benjamin Schreer, "Towards a 'Post-American' Alliance? NATO Burden-Sharing after Libya," *International Affairs* 88, no. 2 (2012): 314.

17. James L. Cook, "Military Alliances in the 21st Century: Still Relevant after All These Years?" *Orbis,* Fall 2013, 568.

18. Alejandro P. Briceno, "The Use of Cultural Studies in Military Operations: A Model for Assessing Values-Based Differences," in *Applications in Operational Culture: Perspectives from the Field,* ed. Paula Holmes-Eber, Patrice M. Scanlon, and Andrea L. Hamlem (Quantico, VA: Marine Corps University, 2009), 36.

19. Major General Duane A. Gamble and Colonel Michelle M. T. Letcher, "The Three Dimensions of Interoperability for Multinational Training at the JMRC," 6 September 2016, at https://www.army.mil/article/173432/the_three_dimensions_of_interoperability_for_multinational_training_at_the_jmrc.

20. Myron Hura, Gary W. McLeod, Eric V. Larson, James Schneider, Daniel Gonzales, Daniel M. Norton, Jody Jacobs, Kevin M. O'Connell, William Little, Richard Mesic, and Lewis Jamison, *Interoperability: A Continuing Challenge in Coalition Air Operations* (Santa Monica, CA: Rand, 2000), 13.

21. Gunnar Fermann, *Coping with Caveats in Coalition Warfare: An Empirical Research Program* (Cham, Switzerland: Springer International, 2019), 5.

22. Per Marius Frost-Nielsen, "Conditional Commitments: Why States Use Caveats to Reserve Their Efforts in Military Coalition Operations," *Contemporary Security Policy* 38, no. 3 (2017): 372.

23. Mark Erbel and Christopher Kinsey, "Think Again—Supplying War: Reappraising Military Logistics and Its Centrality to Strategy and War," *Journal of Strategic Studies* 41, no. 1 (2018): 519–20.

24. Gregory Alegi, "The Italian Experience: Pivotal and Underestimated," in *Precision and Purpose: Air Power in the Libyan Civil War,* ed. Karl P. Mueller (Santa Monica, CA: RAND, 2015), 206.

25. Roger H. Palin, *Multinational Military Forces: Problems and Prospects* (Abingdon, UK: Routledge, 2013), 43.

26. Maria E. Burczynska, "Multinational Cooperation: Building Capabilities in Small Air Forces," *European Security* 28, no. 1 (2019): 99.

27. Steven Paget, "Interoperability of the Mind: Professional Military Education and the Development of Interoperability," *RUSI Journal* 161, no. 4 (2016): 42.

28. Wyn Rees and Lance Davies, "The Anglo-American Military Relationship: Institutional Rules, Practices, and Narratives," *Contemporary Security Policy* 40, no. 3 (2019): 324.

29. Joseph Soeters and Irina Goldenberg, "Information Sharing in Multinational Security and Military Operations: Why and Why Not? With Whom and with Whom Not?" *Defence Studies* 19, no. 1 (2019): 37–38.

30. Christopher Hubbard, *Australian and US Military Cooperation: Fighting Common Enemies* (Aldershot, UK: Ashgate, 2005), 134.

Contributors

Maria E. Burczynska is lecturer in air power studies at the University of Wolverhampton.

Andrew Conway is senior lecturer for the University of Portsmouth at Royal Air Force College Cranwell.

A. Walter Dorn is professor of defense studies at the Canadian Forces College and the Royal Military College of Canada.

Bert Frandsen retired as associate professor of joint military operations at the Air War College, Maxwell Air Force Base, Alabama, in 2017.

Richard P. Hallion is a trustee of Florida Polytechnic University.

Benjamin S. Lambeth is nonresident senior fellow with the Center for Strategic and Budgetary Assessments.

John Moremon is senior lecturer in defense studies at Massey University, New Zealand.

Steven Paget is the University of Portsmouth's director of academic support services at Royal Air Force College Cranwell.

Matthew Powell is teaching fellow for the University of Portsmouth at Royal Air Force College Cranwell.

Stephen L. Renner recently retired as professor of strategy and security studies at the School of Advanced Air and Space Studies, Maxwell Air Force Base, Alabama.

Corbin Williamson is assistant professor of strategy at the Air War College, Maxwell Air Force Base, Alabama.

Index

A MITCHELL INSTITUTE BOOK
AVIATION AND AIR POWER
Series Editor: Brian D. Laslie

In his work *Winged Defense,* Brigadier General William "Billy" Mitchell stated, "Air power may be defined as the ability to do something in the air." Since Mitchell made this statement, the definition of air power has been contested and argued about by those on the ground, those in the air, academics, industrialists, and politicians.

Each volume of the Aviation and Air Power series seeks to expand our understanding of Mitchell's broad definition by bringing together leading historians, fliers, and scholars in the fields of military history, aviation, air power history, and other disciplines in the hope of providing a fuller picture of just what air power accomplishes.

This series offers an expansive look at tactical aerial combat, operational air warfare, and strategic air theory. It explores campaigns from the First World War through modern air operations, along with the heritage, technology, culture, and human element particular to the air arm. In addition, this series considers the perspectives of leaders in the US Army, Navy, Marine Corps, and Air Force, as well as their counterparts in other nations and their approaches to the history and study of doing something in the air.